T0227418

Going Mobile

BUILDING THE REAL-TIME ENTERPRISE
WITH MOBILE APPLICATIONS
THAT WORK

by **Keri Hayes**
&
Susan Kuchinskas

CRC Press
Taylor & Francis Group
Boca Raton London New York

CRC Press is an imprint of the
Taylor & Francis Group, an **informa** business

CRC Press
Taylor & Francis Group
6000 Broken Sound Parkway NW, Suite 300
Boca Raton, FL 33487-2742

First issued in hardback 2017

ISBN 13: 978-1-138-41236-1 (hbk)
ISBN 13: 978-1-57820-300-0 (pbk)

⊙ TABLE OF CONTENTS

⊛ FOREWORD

Mobile Means Business

There are three fundamental realities affecting every organization today.

First, the pace of business competition has noticeably quickened in recent years, placing greater demand on fast decision-making that is based on direct access to information emanating from all corners of the business. Business leaders cannot tolerate late, incomplete, inaccurate nor frankly un-auditable information given the economic reality and business scrutiny of the last year.

Despite this fundamental need, companies are overwhelmed with information. Few companies or government agencies lack access to the information they need to address critical issues. Rather, the challenge they face is getting to the exact information at the time they need it.

This second point is complicated by the third reality that workforces have become increasingly mobile. Fewer people are bound to a desk from 9-to-5, and whether they are at a customer site 500 miles away from the office or in a conference room down the hall from their desks, they need timely access to business applications and information to effectively get their work done.

Going Mobile presents a roadmap for how organizations can simultaneously address these three critical realities and leverage wireless data as an enabler of productivity gains to create new business value and competitive advantage. These benefits can be achieved by making wireless a core component of an enterprise's business strategy.

✪ BENEFITS OF WIRELESS AS A BUSINESS TOOL ARE REAL

Our job as managers is to look for ways to reduce operating costs while systematically increasing customer service and loyalty. Wireless solutions can help companies excel in these areas by producing critical information on demand. By plugging into corporate repositories, these solutions break down the barriers to information access, allowing mobile workers to get business done wherever they may be.

Wireless solutions present the opportunity to reshape business models, bringing new information to mobile workers and redefining standard work processes. Take the example of RotoRooter, a plumbing company whose goal is to respond as quickly as possible to emergencies. For RotoRooter, knowing the location, status and skill level of each of their 4500 plumbers is imperative to obtaining and retaining satisfied customers, and staying a step ahead of the competition. Traditionally, work schedules have been dispatched once each morning, with status updates and schedule changes relying solely on voice communication and the tracking system requiring manual updates.

To streamline its data management processes, RotoRooter deployed a wireless data solution comprised of a rugged, GPS-enabled phone linked wirelessly to an automated dispatch and trouble-management system. This cost-effective solution allowed RotoRooter to improve routing efficiency, which, in turn, shortened emergency response times, increased average visits per day and enhanced data accuracy and visibility. ROI was evident in less than six months.

The benefits realized by RotoRooter translate to almost any industry with a mobile workforce. With wireless access to data and work orders, transportation and delivery companies can track assets in real time and rationalize the number of separate trips drivers make each day. In addition, construction companies can increase job schedule accuracy and material delivery, helping finish projects faster and at a lesser cost.

✪ CONSIDERATIONS IN DEPLOYING A WIRELESS DATA SOLUTION

The foundation for the deployment of wireless data applications is a mobile infrastructure that not only supports today's business processes but can also adapt as new applications emerge. This infrastructure goes beyond the network provider to also include an ecosystem of device manufacturers, enterprise software companies, platform vendors, wireless application developers and systems integrators.

Deciphering which partners best meet a company's needs is not simple and poses a unique challenge for decision makers. As companies consider wireless data solutions, the components they should look for include:

Wireless Network. Deployment of wireless business solutions demands first and foremost a reliable, secure and scaleable network. When reviewing wireless networks, companies should examine consistency, ease of provisioning and use, breadth of available applications, service level support and the range of security solutions offered.

Enhanced Wireless Services. A mobile business environment needs more than data-only wireless business applications. For example, the instant voice capabilities of push-to-talk provide long-range connections with co-workers, clients and partners, providing a valuable and complementary service to address business and customer service issues quickly.

Management and Development Tools. Companies must determine how they will develop and remotely manage the applications used by their mobile workforce while considering variables such as application distribution, authentication, remote access and data management. Fortunately, there are simple yet sophisticated tools available that support the creation of mobile applications, provide secure download of applications and offer the ability to manage private libraries of applications on wireless network sites.

Support. The network, applications, deployment and integration a company chooses to meet its wireless needs are insignificant without the backing of an experienced and specialized support system. Companies should look for wireless solutions from providers whose proven reliable assistance is a benefit to their customers.

✪ APPLICATIONS ARE CRITICAL COMPONENTS

The process of choosing the network, service, tools and support required for a mobile workforce ultimately substantiates its success in the ability of a company to execute productive and valuable business applications and demonstrate ROI. Companies can choose from an array of turnkey applications, from mobile office solutions such as wireless email and two-way messaging, to fleet management, field sales support and project management. Notably, today's market also is seeing a growing portfolio of vertical applications targeting needs of industries such as transportation and logistics, manufacturing, construction, public service, utilities, healthcare, real estate, and financial and professional services. *Going Mobile* takes a close look at a number of user applications through case studies that provide a practical view at how wireless data is being successfully deployed today.

✪ THE TIME IS NOW

As *Going Mobile* examines, wireless data applications can be designed to scale from workgroup to enterprise-wide deployments, facilitating true workforce mobility. And they are here now—already being deployed successfully by companies large and small. Today's wireless data solutions support multiple wireless device types and applications on secure, scaleable networks. Still emerging are integrated voice and data services which run over both local area 802.11 wireless LANs, wide area networks including 3G and even personal area networks through technologies such as Bluetooth.

There's no question that businesses see the benefits of an increasingly mobile workforce, particularly when employees—both in the field and in the office—are armed with the right wireless network, devices, applications and support. The type of increased productivity companies can achieve with wireless data ranges from 50 to 90 percent in back office support for field service applications, and from 20 to 40 percent in lower costs through better visibility into inventory systems and controls, and increased efficiency in inventory management.

Going Mobile provides an excellent starting point to unlocking the productivity gains that can be achieved with wireless business solutions.

—Greg Santoro

Greg Santoro serves as vice president, Internet and Wireless Services, for Nextel Communications, Inc. Mr. Santoro is responsible for product management and business development for Nextel's suite of voice and data products and services. He also manages Nextel's e-business services organization, which is responsible for strategy, design and implementation of Nextel's Internet-based sales and customer care services delivered through www.nextel.com.

✪ ACKNOWLEDGMENTS

We would like to acknowledge all the great insight, opinions and information garnered from our conversations with mobile vendors, users and innovators.

Thanks to the individuals from the following organizations who took the time to lend us their thoughts:

3Com, Air2Web, @Road, ADC Telecommunications, Advanced Technology Ventures, Allied Business Intelligence, Associated Food Stores, AvantGo, Axeda Systems, Bearing Point, Bermai Incorporated, BP Tanker, Captaris, Computer Associates, Concourse Communications, Datac Control International, DataRemote, Dr Pepper/7 Up, EnvoyWorldWide, First Service Networks, Gartner, GeoSpatial Technologies, iAnywhere Solutions, iConverse, InServe, Intermec, iPass, IPWireless, LG Electronics, Long Island Power Authority, Mayday Mayday!, MercuryMD, Microsol, Military Sealift Command, Mockler Beverage Company, Montana Rail Link, National Law Enforcement and Corrections Technology Center, Northeast Medical Center, Numerex, Optimus Solutions, OracleMobile, Ovum, ParTech, PeopleNet, Pharmerica, PepsiAmericas, Petro-Chemical Transport, PowerLOC Technologies, Qualcomm, Rainbow Mykotronx, San Mateo Police Dept., Skycross, Skyline Chili, Solid Technology, South County Fire Protection Authority, Southern California Edison, Southern California Water Company, Southern LINC, Summary

Systems, Taylor-Made Adidas Golf, Telecommunications Engineering Associates, Thinque, Ubiquio, Unity Software Systems, ViaFone, Virginia Task Force 2, Wavelink, Wherenet, Xora, XcelleNet.

We would also like to thank Brian McDonough for his research assistance.

Chapter 1

Envisioning the Real-time Enterprise

Cell phones are everywhere, turning us into always-available, anytime, anywhere communicators. PDAs (personal digital assistants), those cool handheld devices that let you check your calendar or note an appointment with a few taps of the stylus, are almost as commonplace these days. But for business, it's not about the gadget. What's more important to the business world is the way mobile devices can extend enterprise information to a wider, more dispersed group of workers.

There's a pervasive shift currently taking place in terms of how companies look at business processes, and the ubiquity of mobile devices and wireless networks play a key role in making this new vision of business a reality. The issue is speed. Analysts at research firm Gartner have started talking about the "real-time enterprise," a leaner, faster business model in which an emphasis on immediacy—and the technology to enable it—eliminates costly lag times for everything from research and design to customer service.

Central to the real-time model is having access to data virtually at all times, whether you're at your desktop or not. And key to that, of course, is technology that gives workers the same desktop data whether they're in the field, in a warehouse, on a service call, at a conference or working from home.

Gartner cites several advantages to the real-time enterprise. At the operational level, these include reduced inventory and better customer service. For managers, it's flexibility: You can move more quickly to take advantage of opportunities and minimize the damage by being able to shift course when problems arise. At the leadership level, strategies to cope with and exploit changes in the marketplace and larger economy can be implemented more rapidly.

The trend is easy to see. If this book had been written 20 years ago, the question would have been: "Is a computer for you?" A hundred years ago, it would have been: "Would a telephone help your business?" Remember when you had to allow six to eight weeks for anything ordered by mail to show up on your doorstep? Today, purchases made online can arrive overnight if you spend a few extra bucks on priority shipping. Software can be downloaded immediately. The slightly cumbersome fax machine cut several days off the time it took to send a document cross-country, then email cut that to a few seconds. Already it's common to bring a laptop with PowerPoint presentations or other data to a sales call. How much longer will it be before being unable to answer a potential client's question because the most detailed product information or most recent inventory count is back at the office will not only cost you the order, but get you blank stares of disbelief?

The benefits of extending enterprise applications and data to workers on the go applies not only to the mobile workers themselves, but to the corporation as a whole. It would be a mistake to look at a sales professional with a wireless order placement application and see only the time he saves. The real benefit is in the back office, where no one has to key in the order once the sales rep writes it down and faxes it in or drops it off. The real benefit is in the seamless integration between order placement and invoicing/billing systems. The real benefit is having real-time, digital sales records available for company analysis and for customer service.

For some companies, mobile computing allows them to leapfrog the sometimes painful e-business stage entirely. There are plenty of stand-alone mobile applications that don't need complicated backend systems to function. At the same time, workers who have never used a computer can quickly learn how to operate a hand-held device. Some industries, including construction, home health care and trucking, are using mobile applications to increase the productivity and work processes of workers who used to lug around paper. Such companies can benefit from the automation and accessibility provided by mobile devices without the expense of installing desktop computers, servers, local area networks and complicated enterprise planning software.

When technology changes the world, the world has to keep pace. Businesses that don't adapt will die. That's not to say you should immediately figure out as many ways as you can to provide remote access to your business systems. But you must chart your route to the always-connected, always-communicating future.

✪ YOU'RE ALREADY THERE

One question you don't have to answer is whether to take your organization wireless. That happened while IT managers weren't looking, says Michael Doherty, a senior analyst with wireless research firm Ovum. What executive or mobile professional doesn't already have a cell phone, if not a PDA and a laptop? The real question, says Doherty, is, "My company has already gone wireless—how do I add to or leverage that?"

Most mobile business users rely on voice rather than data. And nearly everyone who uses a cell phone for business is expensing all or part of the cost. According to Gartner, in late 2002 the enterprise was claiming 30 percent of employees' mobile costs and subsidizing most of employees' wireless voice services.

That's where the losses from poor planning start. Picture the midsize firm that's paying the cell bills for a dozen employees or more, at rates the employees "negotiated" with all the leverage of a random personal customer. They picked plans the carrier offered, and if they knew the office would pick up the bill, they might not even have taken the most cost-effective plan open to them. So a company is left with multiple carriers, multiple plans, and multiple waste, when negotiating with a single carrier for the best business rate on multiple phones and a single plan for bulk-rate airtime could provide billing simplicity and real savings.

Mobile data services, whether wireless or synch-and-go, have not yet achieved the high penetration of voice, and that's where enterprises are lucky—there's still time for them to take control. And seize control, they must. Data services are harder to integrate and have a bigger impact on the corporate IT infrastructure. Whereas mismanaged voice use just wastes some money, the emerging mishmash of data services can cause real chaos.

Workers who want to use PDAs are already buying their own and setting up synch stations, dialup access, and, in some cases, even installing their own wireless hubs in their offices or cubicles. Not only do you have immediate security risks, but down the road you're going to find none of your employees have cobbled together matching solutions, tripling your support headaches and hobbling any belated attempt to standardize.

To meet the challenges of doing real-time business and integrating the technologies that will support it, Gartner estimates that enterprises will have to increase their IT budgets by 10 to 15 percent per year. To get the most for that money, the company must do a careful analysis of which mobile applications will pay off and how they'll improve individual productivity and business processes, and then create a plan for testing and incremental implementation of technology.

✪ MOBILE OR WIRELESS— WHAT'S THE DIFFERENCE?

Before facing these challenges, you'll need a clear understanding of what mobile and wireless technologies entail, so that you can best decide how they suit your needs. Mobile is the ability to easily carry a computing or connectivity device from location to location. Wireless is the ability to connect to remote servers to access information and applications over a wireless network.

Not all mobile devices are wireless—sometimes a worker just needs to carry information or applications with her on the job, but doesn't need to connect remotely to servers. Other workers may need to connect to remote servers only a few times a day or once a week. Under these circumstances, your best bet is a mobile device that "synchs" to exchange data. To synch the device, you connect it to the desktop computer via a special cradle, or, in some cases, directly to a modem and telephone line. Both devices then download or upload data to "refresh" the handheld device so that all data and applications between the desktop and device match.

One of the first decisions to make for a proposed mobile data application is whether it should be wirelessly enabled or whether users can rely on synching data at intermittent times. Not every handheld device can connect wirelessly. Of those that do, many are limited as to which network they can use. A wireless modem—whether added on or integral—ups the price of the device; service plans must be paid for separately, increasing the cost. Due to network speeds, wireless access can be so slow in some cases that it's not worthwhile.

Of those companies that do need wireless connectivity, most won't need to make any decision at all about what kind of wireless network their data should travel over. It's a question that usually answers itself, after all the other specifications for the project have been defined. Some applications vendors and integrators have standing agreements with proven connectivity providers so that applications don't have to be tested each time over their networks. (The exception is military and public safety organizations; for a discussion of their special network considerations, see Chapter 8.)

Deciding between mobile and wireless requires analyzing the expected benefits versus the greater cost of the device and the connectivity. We'll talk more about this decision process in Chapter 3. Depending on whether you choose a mobile or wireless solution, there is a plethora of devices available to meet your needs. Here's a list of the hardware available:

* **Laptops:** Laptops were the first mobile computing devices, but by today's standards, they just barely qualify as mobile, because you have to sit down to use them. Still, their utility has been proven—in many companies, laptops instead of desktop machines are issued to employees. Deciding whether to make laptops an integral part of your business is a matter of the cost justifying the extra work time that employees may gain by being able to work in various out-of-office locations such as the airport, public transportation, hotel rooms, conventions, home.

The disadvantages of laptops are that they're expensive, heavy, fragile and sometimes cumbersome for workers in the field. They also require a certain degree of sophistication and training—not every employee may know how to use a computer. For certain employees, a laptop equals information overload: It can be too time-consuming to start up a computer and open a software program to gain access to a key piece of information. There are also security issues with laptops: If you're essentially carrying your entire desktop with you, with all its information and applications, there's more at risk.

Connectivity for laptops: Some information on the laptop, such as email, may be automatically exchanged with corporate systems when the user returns to the office and logs in to the corporate intranet. But for some information, the user must manually synchronize the laptop and the systems' files, for example, by uploading reports to the server. When operations demand updated information while in the field, laptops can go wireless via the cellular networks or via 802.11 networks, which use different transmitter systems and different protocols to connect mobile devices to the Internet. For cellular connectivity, the laptop may be fitted with a cell-phone plug-in adapter. This connection is limited to the 9.6 kilobytes per second at which the current generation of cellular networks can transmit data; this can eat up minutes from the cellular plan and make costs zoom. An internal "air card" lets the laptop access 802.11 networks, which offer fast broadband connection speeds of up to 11 megabytes per second; however, these networks have a small footprint of around 300 feet. Many companies now provide 802.11 networks on their corporate campuses, installing several nodes to provide full coverage of up to several acres. (For a fuller discussion of these wireless local area networks [WLANs], see Chapter 6.) Some consumer-oriented businesses, including coffee shops and airport executive lounges, now offer an "Internet café" model, providing access to their WLANs either free or for an hourly fee.

✳ **PDAs:** While some PDAs are designed for simple personal-information management (PIM) tasks such as keeping an appointment calendar, many offer almost as much performance as a laptop—without the bulk. PDAs are small and light—they easily fit into a pocket or purse. There are two main operating platforms: the Palm OS and the Pocket PC platform. For businesses with mostly white-collar workers, PDAs have many advantages. Palm devices offer quick and easy downloads of thousands of applications, while Pocket PC devices run stripped-down versions of most Windows programs and allow users to surf the Web in full color. With their large screens and point-and-click interfaces, PDAs can be easier for non-computer users to manage. Prices have come down and some serviceable models are now available for as little as $100. They're easily replaced, as well. If one is lost or broken, the user can simply purchase one at a retail store and synch it to the desktop to restore missing files and applications. On the other hand, the very convenience of downloading applications to PDAs can create problems for the enterprise: Users may download insecure applications that could compromise corporate systems or leak company data.

Connectivity for PDAs: Most PDA manufacturers offer a choice of models with or without internal wireless modems. Those without internal modems typically exchange information with corporate information systems by connecting via a cable with a desktop computer. Wireless models are designed to work with specific networks, whether the commercial cellular networks or re-branded networks such as Palm.Net.

✳ **Two-way pagers:** These small handheld devices use a different network than cell phones do, one that's "always on." Always-on devices receive information immediately, without the user having to initiate a cellular or Internet connection, so that, for example, an email message can be pushed to the device as soon as it hits the mail server. Two-way pagers usually have tiny keyboards, an input method that some users find faster and less error-prone than writing with a stylus or pointing and clicking. Prices range from the bottom range for handheld devices up to nearly the cost of a Pocket PC device.

In order to run applications on pagers, however, the enterprise must invest in special servers and software. Lower-end models are mobile but not wireless; when the user puts the device into a cradle that's connected to the desktop computer, a button directs it to synchronize data in the device with that on the computer. Because this synching process requires that the worker take the device to the desktop, it can lead to delays in getting information into corporate systems.

Connectivity for pagers: True pagers run on proprietary national networks optimized for data, not voice. The connectivity plan comes bundled with the device.

✳ **Cell phones:** While the use of cell phones to make voice calls is well-entrenched in the business world, their use to transmit data is relatively new, and the kinks are still being worked out. Still, cell phones offer some real advantages as wireless data devices. First, they're relatively inexpensive and highly portable. Second, it's likely that workers will be familiar with their basic operation, which can greatly reduce the need for training. Finally, they are designed first and foremost to make use of a wireless connection. On the other hand, their small screen size greatly limits the amount of information that can be displayed on a single screen, making them difficult to use for applications that require, for example, forms with multiple fields. Because users must use the keypad for data entry, input of information can be prohibitively slow.

Applications may be accessed over a wireless Internet connection using Wireless Application Protocol (WAP) or, on some newer phones, downloaded as Java or BREW applications. Cell phone/PDA hybrids—devices that attempt to combine the best of a cell phone and a PDA—are similar in size and shape to a PDA, but include a telephone keypad as well as either a touch screen or a small keypad. These eliminate the need for workers to carry both a cell phone and handheld device. From the phone user's standpoint, they're a little more cumbersome; the ideal form factor for these "hybrids" is still evolving, and many of the models seem to make placing regular voice calls more cumbersome. Their higher price may outweigh the advantage of greater usability and make them less cost-effective for corporate use.

Connectivity for cell phones: Handsets continue to be carrier-specific; however, most national cellular service providers have begun to roll out next-generation or 2.5G networks that combine two channels, one for voice and one for data. Within the next few years, the carriers will upgrade to all-IP-based networks. In the meantime, pricing and availability of 2G networks remains extremely variable and changes frequently.

✳ **Industrial devices:** Many businesses have special needs that can't be satisfied with the mobile devices that were born as personal information management tools. For example, the construction industry often purchases ruggedized devices with tough shells designed to withstand getting knocked about, while workers who formerly used forms on clipboards find a tablet-shaped device best suits their needs. Options include handheld, vehicle-mounted, and even wearable devices. Input can be via a keypad, touch screen, scanner or some combination of those.

The big advantage of this class of devices is that there are many more options available. Unlike devices designed for consumers, these often have a choice of operating systems and connectivity. They may be dual-band, connecting to both the cellular networks and wireless LANs. There's a great variety of off-the-shelf applications designed for vertical markets, and providers of these devices usually offer software development and customization, as well. All these advantages come with a higher price tag, however, and there can be a longer lead time for delivery if there's customization to be done. Unlike off-the-shelf devices, if an industrial device breaks, gets lost, or is stolen, replacements aren't available at your local office-supply store. As the cost of retail PDAs drops, some companies have found that it's more cost-effective to use and abuse consumer Palm or Pocket PC devices, then buy new ones if they break, rather than outfitting the workforce with specialized devices.

Connectivity for industrial devices: Most use the synch method for transferring data. Wireless connections may be to wireless local area networks or to commercial networks. Connectivity is typically handled by the vendor.

Table 1.1 Many Kinds of Mobile Devices

Device	Description	Vendor
Tablet PCs	A fully equipped personal computer that also allows the user to take notes using natural handwriting, tablet PCs allow the same type of wireless access that laptop PCs offer.	Fujitsu, ViewSonic, DT Research
Palm OS handhelds	Also referred to as personal digital assistants (PDAs), a handheld can be conveniently stored in a pocket or purse. Equipped with a wireless modem, handhelds can access information via cellular networks from Web sites, including many company applications, email, and instant messaging.	Palm, Hand-spring, SONY, Symbol, 3Com, NEC, Compaq
Pocket PC handhelds		HP, Compaq Casio, RThere, Intermec
PC handhelds		HP, Casio, NEX, Sharp
Email Pagers	Wireless by design, a two-way pager allows you to send text messages and data as well as receive it. It often serves as an alternative to a cellular telephone.	Motorola, RIM Paging, Net-work, Mobile Media, Air-Touch, Sharp, Apple, HP
SMS-enabled phones	SMS (Short Message Service) is a service for sending messages to mobile phones that use Global System for Mobile (GSM) communi-catrion. SMS technology does not require the mobile phone to be active and within range for a message to be received. Messages can be stored until the phone is active, and sent once it is within range (e.g., Sprint text messaging).	Ericsson, Motorola, Samsung, Nokia, Sprint, etc.
WAP-enabled phones	WAP (Wireless Application Protocol)-enabled cellular telephones can be used to access information on the internet, including email, Web sites, newsgroups, and instant messag-ing over existing cellular networks.	Ericsson, Motorola, Samsung, Nokia, Phone.com

Table 1.1 Many Kinds of Mobile Devices *(continued)*

Device	Description	Vendor
Palm OS smartphones	Smartphones are a combination cellular phone and PDA that access and transmit data over existing cellular networks. In addition to functioning as an ordinary cellular phone, a smartphone's features may include: Wireless email, internet Web browsing, fax, personal information management, LAN connectivity, data entry, local data transfer between phone and computers, remote data transfer, and remote control of home or business electronic systems.	Kyocera, Samsung, etc.
Stringer smartphones		Not available yet

Source: BearingPoint

⊛ WHY YOU NEED THIS BOOK

Pressed on one side by eager technology vendors extolling the wonders of the real-time enterprise and on the other by techno-fanatic staffers who crave enterprise support for their ultra-cool new gadgets, there are few executives who don't need to take a stance on mobile data. Early hype about m-business, m-commerce and m-customers was followed by too few success stories to make corporations comfortable charging into wireless, but the question isn't going to go away. The dawn of 2003 saw the launch of a welter of snazzy new devices supported by lavish consumer ad campaigns touting mobile as a lifestyle.

If they're not already, everyone from CEOs, CTOs and CIOs down to business strategists, entrepreneurs, small business owners, IT managers and business unit executives will be besieged by customers, staff and vendors begging them to get mobile pronto. These execs need to be able to understand the bewildering variety of options and approaches fighting for a place in the briefcase or the pocket, and to be able to counter both sales pitches and skepticism with a clearly articulated vision.

The mobile question reaches every kind of industry and every type of business, from the small real estate office to the global communications company, from a local plumbing contractor to a regional hospital. Sales managers must decide whether wireless access to email and customer information will make their sales force more competitive or more scattered. Retail chains must weigh the benefits of letting clerks query the inventory database from the sales floor against the expense of providing them with handheld devices.

The potential value of mobile applications is only the first burning question. In order to move forward, these same executives and managers must struggle through a mazelike marketplace fraught with acronyms to understand which vendors and products will deliver true value. They can't allow themselves to be led, they must be informed about everything from network infrastructure to user interfaces so that they can lead the process.

Planning mobile applications demands involvement from many different entities within the enterprise. Top strategists must help determine whether automating mobile workers can enhance productivity and make the firm more competitive. Executives responsible for various lines of business must get on board to define how operations will change in response to and in support of mobile access to data and applications. Members of the IT department must be part of the planning team because IT will not only have to support mobile users, it will be charged with making sure that remote access to corporate systems doesn't compromise their operation or security.

There's no denying the trend toward worker mobility. Cellular telephones freed many kinds of workers from having to be at their desks to communicate with customers and co-workers, creating a mindset that privileges instant communications and constant availability. According to research firm In-Stat/MDR, about 20 percent of U.S. workers could be called mobile, with the need to access corporate data and systems while on the go. In-Stat/MDR expects the number of business users of wireless to grow from 6.6 million in 2001 to more than 39 million in 2006.

This instant and constant availability of the worker breeds the expectation of equal access to whatever information or tools that worker uses for the job. When a customer knows that her email reached the guy who services her printer, it's hard for her to accept that he won't be able to answer her question until he gets back to his desk.

○ WHAT YOU'LL LEARN FROM THIS BOOK

We aim to provide an objective analysis of the opportunities for working harder and better while saving time and money by using mobile and wireless applications. We'll arm you with the information you need to ask hard questions of the mobile vendors who come knocking on your door and to form a strategic vision and tactical plan for incorporating real-time/anytime communications into the business.

In Part One, we'll map the decision process necessary to proceed from developing a strategic vision for mobile data in the enterprise down to tactical questions such as what kinds of vendors to work with. In Part Two, we offer a market-focused view of wireless, broken into the categories of mobile/wireless implementations. Within each category, we'll provide an overview explaining how a particular service is being used by various industries, or how an industry is leading the way in using mobile or wireless applications. Then we'll provide at least two relevant case studies that demonstrate how a particular company or agency has saved time, money, or labor by implementing the project.

When you've finished this book, you'll understand the vocabulary and concepts of the mobile milieu. You'll know whether mobile is right for your company and whether it's right right now. If you do decide to move forward, you'll be armed with a battle plan that will help you become your company's mobile hero.

Chapter 2

Is Mobile for You?

The argument you'll hear a lot from vendors is that wireless and mobile technologies are going to come after you, if you don't go after them first. But even if you accept the premise that most businesses have to consider some kind of mobile or wireless implementation to keep pace in the 21st century, that doesn't mean you need it today. Nor does it mean that you need to give everyone from accounting down to the mail room a smart phone, PDA or wireless Ethernet connection in their cubicle. The real question is, is mobile for you *right now*?

It's an answer you must get right. Research firm Gartner likes to talk about the real-time enterprise, a business that eliminates the lag time in receiving critical information and acting on it. While championing this just-in-time information gathering and the technologies that will enable it, Gartner also warns that by 2007, only half the workers given mobile data will actually reap any benefit from it. That's a pretty poor return on investment, which the analyst firm says will be attributable to poor implementation and management, plus the high costs involved. That's why careful management and wise selection of the right technologies to do the trick are more essential than ever.

"When deciding on which applications to support wireless or mobile, avoid the hype about what is popular or what is the best application of the moment," says Gartner research director Phil Redman. "Address the pain points in your company, where real-time or near-real-time information can ease the pain. There is no one best enterprise application. It will take multiple applications to provide value and a business case that covers both the hard and soft returns."

Early mobile applications tended toward the PIM—personal information management in the form of email, calendar and address book. The heroic mobile project will be specific—not "all the time, everywhere, access to everything," but, "We made this easier." Some users need constant access to a steady stream of email. Some don't need to be updated that often, but when they need data, they need it big, with charts and color graphics. To create a successful mobile application, you must segment your potential users not by job title or division, but by the type and immediacy of data they need.

For the enterprise, it shouldn't be about the gadgets. Looking at a potential expense and essentially daring it to be worthwhile pretty much guarantees a failed project. On the other hand, a lot of corporate decision makers have tended to classify the latest technological innovations in the early adopter category—frivolous toys that can't promise a decent return on investment. They may be right in the specific, but, again, are using the wrong approach—starting with the technology.

Instead of sorting through the welter of devices, applications and vendors in the marketplace, start by examining your current business processes and how they could be improved. How do your workers use data? Through email, spreadsheets, a database, a Web site? Would they be able to get more done, and in less time, if that data were not tied to the desktop? Are there times when the answers to questions employees ask—or need to answer—are not in the same place as the employees who are asking the questions?

For example, do sales people in the field have to admit to a potential customer that they don't have certain information at their fingertips? They may promise an email in the morning, or a fax when they get back to the office next week. Meanwhile, an opportunity to hold a client's attention, to make an impression, to close a sale, has passed.

An inventory clerk goes back and forth from desktop computer to warehouse aisles. Data entry is being duplicated—logged on a clipboard, perhaps—then entered by hand into a computer. Lag time and human error result in incorrect counts. You're selling things you don't have in stock or buying things that are already falling off your shelves.

Repair personnel making service calls can't answer customer questions about the service package, the availability of a needed part or when a follow-up service call can be scheduled. The customer will have to spend half an hour on hold waiting for your jammed call center, where it will take longer to get the phone representative up to speed on the problem than it would have taken the field agent to solve it—if the technology had been there.

In a world in which service expectations increase and patience dwindles and where lean operating budgets demand that more be done with less, a company's data infrastructure must support people on the move. Mobile and wireless apps can make the difference —if they're designed right and used efficiently.

✪ FINDING THE HEROIC MOBILE APP

Any successful technology project must fill unmet needs. Enterprise mobile projects should focus on meeting the real needs of its users; otherwise, it will be used with indifference or distaste. Workers who spend too much time and attention on rote tasks or duplicating work can feel real pain. If you can design a mobile application that releases them from repetitive or low-level tasks, you'll be a hero.

You also need to know your limitations. An agricultural insurance firm that sends agents out to rural areas is not likely to get high-bandwidth networks that will allow them to download large files conveniently. For that matter, there might not even be basic cell service, so mobile synch solutions may be the best option. If the agents' operational areas are well covered by basic wireless carrier-wide area networks, smaller applications such as remote email access might be viable.

Any new technology or work routine takes time to implement. However, if the payoff doesn't seem sufficient, the worker will never get over the adoption hump. Think of how many of your workers with standard desktop PCs don't use the calendar feature in their Outlook or Lotus Notes program. Fortunately, you didn't pay extra for that feature. But if you're going to start paying for new devices and new services, make sure they're relevant to your users and give them only to the users for whom they're relevant.

"Understanding user needs is extremely important, and enterprises should gather information from users, division managers and current operator relationships to institute an adoption and usage policy," says Gartner's Redman. "Those companies that do can generally see a 20 percent decrease in telecommunications costs."

✪ SHOW ME THE MONEY

Between economic hard times and the ongoing drive to become leaner, more efficient organizations, the need to demonstrate clear return on investment is greater than ever. If you can't identify how and when you can get a return on the money you'll spend, mobile is not for you. Despite the plethora of mobile applications that are available, and the substantial number of companies that have implemented at least basic mobile or wireless extensions to information, there is very little firm data on how much of a return on the investment such deployments will offer or on when they'll do so. Vendors are working hard to help their clients find answers to the ROI question; unfortunately for mobile strategists, those that may have found ROI don't want to share that information.

Once you understand your needs, you can estimate potential gains to determine how aggressively to pursue your mobile options. Do put your vendors on the spot. Ask them to provide concrete breakdowns of ROI, which should be well-explained and plausible. The vendors' business case should be similar to your internal assessments.

Mobile offers the potential for a return on investment in many different ways. Consider the following:

* **Saving workers time:** Automating tasks such as inventory or reports may save an hour or more per day. However, this saving doesn't necessarily translate to improved productivity; workers may simply do the remaining tasks slower, or they may not have additional responsibilities that can fill the newly available time.

* **Making workers more productive:** Employees that are armed with better tools and information may be able to do a better job. For example, when delivery drivers know in advance which

customers on their route don't need service that day, they can focus on stops that do. Remote monitoring of vending machines or real-time reporting of deliveries can make drivers much more efficient—but only if the logistics operation is prepared to act on this information.

✳ **Capturing more market share through improved customer access:** For companies whose customers can benefit from closer to real-time information about inventory, pricing, order status or shipping, allowing customers to access this information via wireless devices or enabling them to deliver information or requests to a salesperson on the move could provide competitive advantage. There must be a compelling reason why a voice call to the salesperson's cell phone or a customer self-service portal on the Web are not enough. Your company may have historical data that can help with this decision point. Did establishing your Web site years ago, or some other added automation, result in measurable improvement in sales or market share? Let the past inform your predictions.

✳ **Paperwork reduction:** While eliminating the need for workers to sit at the office keying in notes or reports of activities in the field could make them more productive, reducing such paperwork may save even more time for operational support personnel—the workers who process the forms, re-key them for entry into disparate corporate systems and maintain the paper files. Companies that generate a lot of paperwork may be losing money simply by paying for the otherwise useless space to store them.

✳ **Better inventory management:** Would a wireless or mobile application improve your ordering? Can you calculate the losses resulting from being unable to fulfill sales orders, or to complete service orders because a replacement part isn't available when needed? How about excess inventory? How much did you spend last quarter on products or materials that sat unused on the shelves?

✳ **Improved operations through better information**: While the typical mobile application seeks to better connect remote workers with information and applications on the home servers, the big—and unexpected—win often comes in the form of the data stream that begins to flow back to the home office. When employees begin to file reports and logs electronically, it becomes possible to feed their information into corporate databases where it's available to analytics and e-business applications. Suddenly, the company begins to identify trends that can help it further increase efficiencies, predict problems or take advantage of new opportunities. For example, when field service workers at a utility company begin to log meter readings electronically, the utility can compare readings to maintenance problems and identify equipment that may be about to go down.

✪ DO YOU HAVE EXTERNAL RESOURCES?

Launching a mobile or wireless initiative takes not only money, but labor—and plenty of it. If you decide to build, buy or integrate a mobile application using internal resources, you'll need to identify the required skill sets and the staffers who can handle the task. Some smaller organizations have handled their own mobile projects completely in-house using the Palm platform. If your company does not have a full-blown e-business infrastructure, or if you're considering mobile but not wireless applications, or if you have a relatively small number of users and have the technological aptitude to oversee the project, it's perfectly possible to deploy—and even develop—your own mobile applications. (For examples of organizations that have gone this route, see the case studies "Railroad

Tracks Crossing Gates" in Chapter 7 and "Shipboard Inspectors Carry PDAs From Bow to Stern" in Chapter 8.)

Larger organizations contemplating letting mobile devices tap into their highly developed corporate information systems will need to muster an array of partners and vendors. If the enterprise has strong information technology partnerships and thriving relationships, it may be able to depend on its existing partners and vendors to take charge.

"To find service providers, first, ask a friend," says Michael Doherty, an analyst with wireless research firm Ovum. Existing vendors have a head start on understanding your current business environment and your strategic plan for the future. You've already developed a relationship of trust that can ease the tensions of making the complex and crucial decisions you'll face.

Another reason to begin with your little black book is that you need not only an integrated set of wireless solutions, but wireless solutions that integrate with your existing infrastructure. When faced with hopeless legacy systems or copious amounts of capital to burn, you may opt to redo your wired infrastructure to accommodate your wireless plans. More likely, you will hope to keep your current infrastructure and planning relatively unchanged. It's important to make sure that the wireless functionality you want will play well with the wired applications, hardware and databases you've already set up.

It's not unlikely, however, that the company will need to forge new business relationships with unknown vendors, a process that can be perplexing and time-consuming. To make it harder, there is no true one-stop shop for mobile applications, despite vendor claims of "total solutions." Nor is there a clearly defined hierarchy of vendors. The U.S. mobile network operators have launched divisions dedicated to developing, testing, and selling enterprise applications as part of their business services. At the same time, they're working with small and large applications developers and professional services organizations to help them tailor their mobile data offerings to the enterprise.

Traditional enterprise software manufacturers such as Seibel Systems and Oracle have mobilized many of their applications and sell them directly into the enterprise or through partners. Consultants and integrators have their own relationships with software companies as well as the ability to develop and deploy custom applications. At the bottom of the food chain are small software shops that may focus on specific applications for vertical markets or may have a single application to offer. The burden is on the enterprise to identify its best case for mobile apps, then to vet current and potential partners.

Where once everyone from individual applications providers to major wireless carriers were promising the world and delivering barren islands of functionality, they're now attuned to corporations' sharp focus on returns. This means that vendors know they're expected to be able to demonstrate where the measurable value is before they're going to get a contract.

Even software and platform vendors will admit that it's a buyer's market. There are too many vendors, too many consultants, too many applications and a confusing array of connectivity options. The rare "referenceable" customer is worth its weight in free software and services. Businesses actively in the market for mobile applications and services can often get deep discounts and freebies, especially if they are willing to share their stories or at least give the vendor a hint about the benefits achieved. Pilot projects in which the customer pays little or nothing are the norm in this emerging market, so the enterprise can refuse to buy until it tries.

⊙ WHAT ABOUT YOUR HOME TEAM?

At least one senior member of the IT department should be involved in the project at the strategic level. IT can provide insight from the trenches about what users are really doing, where potential and actual choke points in data management exist, and where any trouble spots are in your system architecture. Without a single mobile vision and an integrated series of solutions, the IT manager will eventually find a mismatched patchwork of applications, each the easiest solution for one worker's immediate problem, but without consideration of an overall, more manageable strategy.

Once you and your outside vendors establish the platform, hardware, applications and the scope of the vendor's involvement and responsibility, it's time to begin marshaling your internal support team. The exact role and responsibilities for IT may become clear only over time and will depend on the types of applications and devices you deploy. Depending on the scope of your mobile deployment, your internal IT staff may be called on to take over the ongoing management of the project. They may order devices and configure them for new employees, then collect devices and shut down accounts for outgoing staff.

Discussions among managers, the IT department, and service and application providers should begin early and occur often. You may need one or more staffers specializing in wireless matters if your deployment is large or complex. Otherwise, mobile simply becomes part of the shared duties of your entire IT department. Smaller organizations may find it worthwhile to provide training on the new applications and devices to the entire IT staff, so that support doesn't become a bottleneck that limits usage of the mobile devices and apps.

Make certain your IT staff will have enough lead time to get up to speed on the mobile project, help roll it out, and provide ongoing user support and maintenance. Inadequate provisioning for ongoing support is one of the classic pitfalls of mobile technology deployments.

✪ TESTING THE VALUE OF A WIRELESS PROJECT

The idea that real-time pressures are going to transform business and make some form of wireless implementation almost inevitable can be a daunting prospect. But don't forget, it's a transformational process, not a revolutionary moment. Companies getting their feet wet with mobile technology have been starting very simply, taking a small subset of users and giving it a single device or limited suite of applications. Replace a few clipboards with PDAs. Test a WiFi environment in the warehouse. Give a few executives WiFi-enabled laptops or PDAs and let them test the high-speed WiFi access available at an increasing number of convention halls, business hotels and airports. Give a salesperson in the field the ability to perform a quick, form-based inventory check. If you have mobile service people, let them synch up their daily schedules and the latest manuals or documents on a mobile device.

To figure out which projects will have value, look short and long. The short view is of that contained test project. Pick something that's likely to have a simple, immediate value and that won't require massive retraining for users. To a great degree, pilot projects can be less concerned with immediate value or potential ROI. If you pilot a wireless project that helps out ten of your 120 employees, the return on investment may not make your CFO weak in the knees, but it may point the way to true savings or efficiencies when it's broadened.

But make sure your tester also has potential as part of a longer-term enterprise approach. The real rewards come not from a wireless app here, a quick synch function there, but through bringing your enterprise to a point at which everything is faster and smarter. Your long-range plan has to be flexible enough to shift as you learn, but you need at least that skeleton vision before you figure out what the first steps are.

Be prepared to take a controlled risk if your analysis shows that mobile might be for you. As you move forward in the planning process, keep your eye on any small returns that have the potential to go big when applied company-wide. Don't roll the dice—and blow your budget—on a mobile hunch, but remember that being a mobile hero for your company includes meeting substantial challenges in the hopes of big operational rewards.

Chapter 3

Defining Your Wireless Project

If you've already read Chapter 2 and decided that a wireless application is right for your business, you've taken the first step. Now comes the hard part—defining what it is you want, need and can realistically accomplish with wireless and mobile applications. Too many companies that skipped this step and rushed into a "brave new world" without assessing the current environment have had unsuccessful deployments.

Many early-stage mobile applications for business were device-driven. For example, delivery service UPS pioneered custom tablets that closely mimicked the clipboards with lists of addresses their drivers used to carry. With the tablets, drivers could automatically log deliveries and capture signatures. Among the general-business white-collar crew, early adopters bought themselves personal digital assistants and cell phones, played with them to find cool—and sometimes useful—things to do, then asked the IT department to help them when they got stuck.

But a device-centric approach doesn't make sense any more—there are too many choices out there. While it's easy to fall in love with a chunky tablet that looks like a Tonka truck with its ruggedized case or a chic platinum rectangle with a tricky flip top—and then figure out how it could be used on the job—there is now such a wide range of mobile devices available, either off the shelf or through applications vendors, that almost any application or task can be supported. Therefore, choosing the actual device on which the mobile application will run becomes almost an afterthought.

Instead, the decision process must be driven by your data and your users. Companies must make the following determinations:

* Which workers have the greatest need for mobile access to data and applications?

* Which tasks can most easily and productively be automated?

* What kinds of data do workers most need when they're away from their desks?

* Which corporate systems, applications, and databases hold needed information that's not easily accessed remotely?

* Does the worker need constantly updated and timely information or only reasonably current information?

* Can the business make use of more data delivered faster from the field?

Decision makers must look especially hard at those last two questions. It's by no means a given that getting or sending more information faster will help a company do business better. In the early days of the Internet, pundits compared finding information on the Web to trying to drink from a fire hose. That same fire hose effect can happen when databases or analytics programs used to receiving batches of reports and consistent data suddenly try to make sense of a gush of real-time data.

GOING THROUGH THE DECISION PROCESS

A step-by-step approach to identifying your needs, potential users, desirable applications and important data will avoid "scope creep"— that frustrating and expensive scenario when the project mushrooms out of control. Here are the questions you must answer, either before you begin to work with vendors or—at least—before you begin to evaluate technology and hardware. To answer them, you'll need to communicate clearly with your managers, employees, colleagues, and partners.

WHO ARE YOUR POTENTIAL USERS?

There are three broad categories of mobile or wireless applications:

* Customer-facing applications
* Partner-facing applications
* Employee-facing applications

The market for customer-facing applications includes such services as remote order placement via a customer's cell phone for restaurants and wireless alert services for customers of retail establishments. These applications have been tested, rolled out, and, for the most part, proven unsuccessful, except from a pure CRM standpoint. There always will be a few customers who find such services useful, but the cost of rolling out these projects often outweighs the benefit.

Partner-facing apps face the same challenges as customer-facing apps: How many of your actual and potential partners are really going to use these services?

Employee-facing applications are what we'll focus on in this book, because this is where most companies are seeing real ROI. In most cases, the real savings for the enterprise market come from an internal deployment. Look at your employees and think about which segments are most likely to benefit from access to enterprise data while they're away from their desks (if they even have desks).

✪ KNOW THY USERS

To understand what your workers need, you'll need to take a fresh look at what they do, and how and where they do it. There are a high percentage of workers who never leave their desks, other than for the occasional staff meeting, at which they need nothing more advanced than a notepad and pen. There tend to be fewer workers who are entirely mobile—out of the office so much that they don't even have a cubicle to call home, and you have trouble remembering their faces. In between are a great swath of workers who might be thought of as semi-mobile. They work at their desks, but are also out and about. They sometimes work from home. They go on sales calls. They spend most of their time on the road, but also come into the office regularly. To assess the range of functions of your workers in your enterprise, ask these key questions:

✳ **Where do they work, and what do they do at different locations?** Identify the several locations where each type of employee does the most productive work. Field-force workers should spend as little time at the office as possible, while a sales or marketing person may spend several consecutive days in the office, then go out to make a presentation. A delivery driver spends the majority of his time in the truck, where he needs to keep track of what's still on the truck, what's been delivered, and what's been picked up.

✳ **What applications do they most often use?** In the office, a marketing person might need access to creative software—creating documents in a word-processing program, working with graphic design on a desktop-publishing program, doing research with a fast Internet connection and a browser. In the field, she might do no creative work or research, needing only to bring along data prepared at her desk and to check her email now and then. Twice a year, she might staff a trade-show booth for several days and want to take much of her desktop data access and connectivity with her.

✳ **What kinds of information do they most need?** Depending on the product, salespeople may need heavily graphical data such as images of products or complex charts detailing performance specifications or other tabular matter. Field-force personnel such as repair/service workers need manuals, which often include charts and diagrams. Sales and managerial personnel may need consistent access to email. Some workers need access to shared files in the corporate database; others are more self-contained. Finding out what information is needed is the first step to figuring out how to provide access to it.

✳ **Which applications call for continuous or frequent updates?** Can some be used offline for a full workday or longer? Certain information, such as open requests for cab service, must be constantly updated via a wireless connection to be effective; others, such as email or daily reports, can wait for synchronization at the end of the day. A worker who makes short trips in the immediate area should be able to update information on the mobile device by synching it to the desktop between trips.

✳ **Whom do your mobile workers inconvenience?** So far, we've discussed mobile applications from the standpoint of making life easier for the worker on the move. But while the manager attending cross-country meetings might not need wireless access for presentations, the dozen workers on her staff who can't move forward on a project until she gives direction are idled because she can't get her email until she returns to work, or until she gets back to her hotel room and plugs her laptop into a phone jack, by which time her staff's workday is lost. After assessing what your mobile workers need, don't forget what their deskbound colleagues, customers and outside partners need from them while they're in motion.

* **What options are feasible in their mobile environments?** You might think that a salesperson needs to be able to wirelessly access a dynamic, complex database at all times in the field, but if you're sending him to rural areas that won't be seeing robust wireless networks for a couple more years—or any cellular coverage at all right now—wireless applications won't do him much good. If your mobile executives spend most of their away time in hub airports, convention centers and business hotels, you may be able to deploy more robust applications to laptops and some PDAs, thanks to the high-speed wireless local area networks (WLANs) becoming increasingly common at those locations. A field-force worker who spends much of her time deep in the bowels of a concrete warehouse could benefit from WLAN access, but not from the spotty wireless connectivity she'd get from a cellular connection.

* **How could mobile data access eliminate unnecessary processes?** Having figured out how your workers currently operate and how to get them the data they need, look for ways to change how they work. Sales- and field-force personnel might make their field calls, then come into the office to file reports

or place orders. Why? Perhaps it would be faster, upon giving them wireless technology, to have them fill out that paperwork electronically in the field, so that data flows more quickly, with less downtime spent driving to and from the campus.

* **Who else could benefit from the deployment?** There may be workers whose needs wouldn't justify rolling out an application, but who would find the app useful if it were available. If wireless email access is necessary for your sales force, would it also be an incremental boon to the executive who routinely spends six hours a day in meetings, away from his desk? Once you've given the functionality to the main user group that truly needs it, it can be far less expensive to extend it to other staff, and that lower additional cost may be justified by the small efficiencies returned. As you gather this information, begin grouping users not by job title, but by mobility needs, to figure out those collateral benefits. Regardless of job title, you may find that you have 20 workers who need remote access to an inventory database and a dozen more for whom it'd be a convenience.

✪ KNOW YOUR OWN "REAL WORLD"

Some managers would answer the above questions by sitting alone in their offices and jotting them down on a notepad. They're probably the reason why Gartner says that up to 50 percent of workers receiving rollouts of mobile applications won't get benefits from them. Heroic mobile strategists don't *tell* their workers what they need, they *ask* them. To get the answers to your real-world questions, employ the following methods:

* **Conduct surveys.** A well-designed survey sent out by email, asking workers easy-to-answer, easy-to-tabulate questions, is a

great start. How much time do they spend away from their desks? Which applications and data would they like to be able to access away from their desks? Do they experience times when they need information that they can't access? How would they like to work differently when they're on the move? What wastes their time?

✳ **Talk to managers.** Managers can be especially useful as skeptics. Sometimes what people say they want, and what they'll actually use, aren't the same. A good manager knows how his staff reacts to change, how aggressive they are about incorporating new efficiencies, and whether they'd use a PDA to synch their calendars or just to download games.

✳ **Assess user attitudes.** Analysts at the Yankee Group report that many enterprises trying to capitalize on mobile applications for sales-force automation have been hindered by slow user adoption. If your staff perceives the new application as a pointless distraction from their tasks or another addition to their workload, they won't use it, and you won't see any benefits. There can be more than one way to solve a problem. Make sure the one you pick is user-friendly. If you do your homework, you should be able to clearly communicate the value of mobile; understanding employees' attitudes will help you sell them.

✳ **Hold focus groups with similar workers.** Taking the opportunity to sit down with groups of employees and chat frankly about what's hard, what works and how it could work better is invaluable. It's important when convening such a group to organize it by the type of work its members do, rather than by department or business unit. Get the sales forces from multiple divisions together to share their experiences, then invite your logistics team, then the delivery drivers. Make sure that everyone in the room has sufficiently similar needs that they can combine perspectives to generate compelling solutions that will benefit the greatest number.

✳ **Find champions.** The worker who's hard-core about technology or aggressive about making the most of opportunity will often introduce under-the-radar hardware or software. Locate workers in your organization who are already using some form of mobile or wireless application. For example, the Yankee Group found that salespeople were leading the charge in adopting wireless and mobile technology on their own, becoming those potential champions in the enterprise. Instead of chastising them for complicating the corporate infrastructure, find out whether their innovations have paid off. What specific efficiencies are they gaining? Ask whether they think you're going in the right direction. If your early adopter isn't excited, you might want to rethink. And if that maverick has shown results, make him or her the poster child for the project.

✳ **Create champions.** It doesn't have to be an early adopter who leads the charge to mobile among the rank and file. An enthusiastic newbie may do even better. In a workforce unused to employing technology, it may be difficult to find individuals who are already technology evangelists. Instead, identify those who are natural leaders, because of their position, their seniority or their social clout. By involving them in fact-finding, planning and testing—and by truly valuing their contributions— you can get their buy-in. When co-workers hear a former Luddite say, "It's easy to learn, and it makes my life easier," that's compelling.

✳ **Envision the device.** Maytag service techs or FedEx drivers making rounds don't need and can't use a cumbersome laptop computer. Is the data you need going to work on a handheld screen? On a smart phone? Will the data display require the larger screen of a laptop or notebook? Will it be used in an environment, such as on sales calls held in corporate conference rooms, that makes larger devices viable?

✪ TECHNOLOGY FOLLOWS FUNCTION

Once you've looked at what you do, and how it could be done better, you're in a position to consider which new technologies can help you get there. Tech isn't the answer to every problem, but when the challenge is to be faster, more flexible and better informed, mobile and wireless solutions can be the answer.

Early wireless applications taking off in the corporate sphere have tended to be very simple—access to email, calendar and scheduling programs, or simple databases. That's due to a number of reasons. For one, it's easiest to adopt simple wireless versions of familiar desktop processes. It takes less vision, user training and capital investment. It's a great way to let wireless and mobile technologies prove themselves within the enterprise. The economic crash that hit the tech sector and general stock market in 2001 didn't help much, either; more elaborate wireless deployments seemed too costly. Finally, when you look at the actual wireless services that have rolled out, there hasn't been a breakout application so powerful or useful that it reaches across vertical markets.

What tasks can be automated?

Once you've determined which users would benefit most from wireless or mobile, think about their daily tasks—talk to them, survey them, and learn what it is that takes the most time and has the possibility of being automated via wireless or mobile.

Wireless or mobile?

This is one of the most important questions you can ask yourself, and it's something we'll come back to again and again throughout the book. (See Chapter 1 for a more thorough discussion of the benefits of each method.) For the purposes of your initial planning, you'll want to think in terms of whether the benefit to your workers lies in having access to real-time information (data that changes and updates by the minute or hour), as opposed to having access to applications or information that's updated daily.

What business processes do you want to integrate with your mobile/wireless system?

How deep do you want your wireless system to go in terms of working with existing systems? Would it be beneficial to have your wireless system integrate with your payroll, invoicing, billing, CRM, or forecasting and analysis systems? Can your existing systems handle the flow of information from a new wireless component and make use of it? If you're not sure, the next step is to get in touch with your current technology vendors and discuss whether they'd be able and willing to work with your new wireless/mobile system.

What changes in your existing technology infrastructure or business processes would you need to make to benefit from a wireless/mobile deployment?

Involving your existing vendors with your new wireless/mobile vendors is key to a successful project, and it's even easier if your software is open source. It's possible that your existing business systems might not be capable of mobile extensions, in which case you'd have to consider an investment in new backend systems to support the new mobile applications.

Your current business processes might keep your company from taking full advantage of mobile extensions. For example, many companies handle accounting by having a person do manual data entry to get information into the system. If employees must take the electronic data from other automated processes and manually re-key it, you may find that you haven't saved enough to justify the project.

What kinds of mobile data do your really want and need?

Given the advances in the wireless technology market, there's a lot of data that a company can collect these days. But a lot of information doesn't necessarily equal a lot of benefits; what you need is a system that gives you critical data without overwhelming you.

Many applications are designed using the principle of exception-based data management. Put simply, if everything is running smoothly, you may not need to know about it. For example, you could fit a manufacturing unit with a wireless connection that would constantly send its operating temperature to your monitoring system and feed the information into a database. This system could provide valuable trend information over time—if you have the ability to analyze it. At the same time, this steady data stream could make it more difficult to actually monitor the machine. Applications that use exception-based rules let you set the parameters of when or under what circumstances you want to know about something system and alert you accordingly.

Will you involve outside vendors, and if so, what kind?

There are a variety of partners you might work with to get a wireless/mobile system up and running. Or you might look at what's out there, say, "No thanks," and decide to build your system on your own. Building in-house has its advantages. Certain costs can be lower, and control over every alteration along the way is in your hands. If you're in a market niche, you may be best able to define the feature set you need.

There are also disadvantages. You might waste time evaluating inappropriate technologies, miss opportunities for efficiencies or find that you've re-invented the wheel. You may not have the technical expertise on-staff—and you may not find that out until it's too late. There are a multitude of vendors and solutions available, and prices are dropping. In this buyer-hungry market, the disadvantages of building mobile applications yourself generally outweigh the advantages.

If possible, look for an outside provider who can handle both the implementation and upkeep. While your in-house IT department may eventually handle technical support for users, unless you create a team whose sole responsibility is the mobile deployment, IT will have neither the expertise nor the resources to evaluate which carrier, which devices, which email application, which WiFi hub and which operating systems should be stitched together into the corporate solution.

How secure does your data need to be?

Wireless security (and the lack thereof) is a hot topic these days, because numerous cases of security lapses have cost companies data, customers and money. Everyone needs a basic level of security, but how much more than that you need depends on what kinds of data you'll handle over wireless connections. Are you transmitting highly confidential customer information over your wireless system? What types of networks are you using? What would the consequences be if your data was intercepted?

What type of wireless network will you use?

This will likely be your last decision, as it depends on your answers to many of the questions above. Tomorrow's data applications will travel over fast networks that, like the Internet, handle all voice and data communications as packets. These true packet-based networks, so-called third generation, or 3G, are "always on," which means that wireless data users wouldn't need to dial up and wait to reach remote servers. Ideally, the servers will be able to deliver information at broadband speeds of up to two megabytes per second, enough to let users transmit bandwidth-hogging music, graphics, and video files.

There would be obvious advantages in 3G networks for businesses that need remote access to schematics, catalogs or photos. The big four U.S. wireless network operators are in the early stages of rolling out their 3G networks. Unfortunately, they're going live more slowly than the operators predicted and often at speeds much less than the maximum. Meanwhile, pricing remains prohibitively high for serious business use or transmitting all but the most critical information. Wireless business applications may run on several different kinds of networks or on a combination of two or more.

✪ BURNING ISSUES

Even all the planning and researching in the world can't take the place of experience. We asked top consultants to dig into some of the thorny questions mobile strategists must confront. Herewith are some best practices that they've identified over the past few years of experimentation.

An Executive-Level Update on Wireless Security
By John Stehman

Wireless security is an increasing concern for enterprises, and unless a coherent solution is implemented that conforms with existing security policies, problems will quickly occur. IT executives should take the time to update themselves on key wireless security issues and then evaluate and select the security measures that are most compatible with existing security policy requirements and the IT infrastructure.

BUSINESS IMPERATIVES

Allowing wireless access to business applications can generate improvements in productivity, and in some cases, produce incremental revenue. IT executives should determine the security levels required for each business application and verify that suitable solutions are available and implemented to meet enterprise needs.

- Security carries a high price tag, depending on the level of protection and the types of security measures required. IT executives should determine the costs to implement, administer and maintain wireless security solutions as part of the IT infrastructure planning process before deploying wireless-enabled applications.

- Security products will continue to improve, and ease-of-use and deployment costs will become more enterprise-budget friendly. IT executives should stay abreast of security improvements on the near horizon, and make certain applications are designed to accommodate new security products as they are released.

Wireless service providers are deploying next-generation (NG) wireless services, and wireless local area networks (WLANs) continue to penetrate the enterprise. However, whether the wireless technology is a WLAN or wireless cellular service, it remains mired in controversy and misunderstanding concerning security capabilities.

Securing enterprise wireless networks requires a balanced blend of security intelligence, security policy, effective user education and continuous monitoring and adjustment. However, Robert Frances Group believes that it will take about three years until wireless services and networks will provide the necessary security enhancements that are business application compatible and can be better assimilated into existing security architectures.

As wired and wireless networks continue to grow and coexist in enterprises, business applications will be required to support user access to both of these technologies. Many corporations already utilize WLANs, and projections indicate they will continue to grow significantly in the coming years.

In many cases, the performance and security requirements for business applications require a level of consistency for both wired and wireless environments. This includes the ability to administer and monitor security performance from a central, integrated point. As a result, IT executives should consider future and present security requirements, as they plan wireless deployments and consider security administration and integration within the IT infrastructure as critical objectives.

The fact remains that IT executives planning to deploy wireless-enabled applications must still contend with limited performance and questionable security capabilities of wireless providers and WLANs. The unsettling disconnect between enterprise requirements and vendor capabilities was obvious to the RFG staff attending the recent Cellular Telecommunications and Internet Association (CTIA) conference. This troublesome situation will compel enterprises to make prudent business decisions that determine if wireless services can provide a value proposition for the enterprise that meets security policy requirements.

Some of the key questions IT executives should explore during the wireless security evaluation and planning process are these:

- Are current wireless security capabilities compatible with enterprise security policy?

- Are business application security requirements, including user devices, network transport and the application, compatible with available security measures?

- What is the potential cost to the enterprise for a security breach and can the breach be quickly detected?

- What is the total cost of ownership (TCO) to implement and administer the required security measures needed to protect an application during its lifespan?

- Can the initial security capabilities be easily upgraded at an acceptable cost during the lifespan of the business application?

- Are the security measures required to protect business applications acceptable to users from an ease-of-use and productivity standpoint?

Another major challenge for wireless security is that it continues to be a work-in-progress with no "one size fits all" solution. In many cases, wireless security is similar to fighting a wildfire—as soon as it appears to be under control, another fire appears at a different and unexpected location. Although there are various point solutions for protecting specific elements in the wireless application delivery path, there is no complete end-to-end solution that is easily managed and administered from a central point via an intuitive graphical user interface.

Likewise, enterprises will discover that robust wireless security requires a tradeoff between the cost of the security products and the level of complexity acceptable to IT staff and end users. Managing and administering wireless security is not easy, and IT executives should take the time to review all potential security products and features to verify that they can be managed and administered effectively. Since most enterprise infrastructures already utilize security products to protect the wired environment, it is important to determine if wireless security products can be integrated, or at least co-exist, with those already installed.

RFG believes it is still difficult to integrate security architectures for wired and wireless application access, and enterprises should carefully evaluate the current architecture and the required wireless security measures to uncover the complexities involved. IT executives should also consider the following questions when planning wireless security:

- Are wireless security requirements different from those currently employed for application access? If so, how?

- Can any existing security products be utilized to provide any or all of the security measures required?

- How will security for both wired and wireless access be coordinated, administered, and monitored?

- Can the required security measures be fully tested in a pilot situation prior to application rollout?

The above questions can provide valuable insight for IT executives who are deciding if wireless access offers a value proposition for specific business applications.

Although IT executives should not be expected to become security experts, they should understand the broad categories of security that could prove beneficial to protecting business applications. A sample of the security solutions and new product areas that should appear on enterprise radar screens are briefly described below:

- **Biometrics**, especially fingerprint and iris scanning, for user authentication at the device and server levels.

- **Virtual private network (VPN)** technology that places a small client footprint on user devices and is validated against VPN server applications.

- **Turnkey security administration solutions** that allow user security templates for different business applications to be created and then automatically applied as required.

- **Security analysis tools** that identify the best solution for different security levels, such as low, medium, and high, and include the ability to identify encryption overhead.

In the next few years, security products will become more intuitive and user-friendly, from both an administration and end-user device standpoint. For example, fingerprint scanners on a PC keyboard could quickly provide reliable user authentication before access rights are granted to communicate with any application server. Security will continue to operate in various layers, but end users will become more detached from today's cumbersome processes that are needed to establish a VPN connection or to complete a "users' rights" validation access from an enterprise authentication server.

IT executives should assume a stronger role to determine and evaluate enterprise wireless security requirements that will protect important applications from hackers and eavesdroppers. Likewise, IT executives should plan wireless security initiatives circumspectly and verify that the available security meets corporate policy and application security requirements. IT executives should also identify the costs and resources required to implement and maintain wireless security, and plan to integrate similar security functions and capabilities in the IT infrastructure whenever possible.

John Stehman is a principal analyst with Robert Frances Group (RFG), an IT advisory service that provides research and consulting to Global 2000 and mid-market executives who require a business-oriented view of the IT marketplace.

The Nitty Gritty: Legal Issues to Consider

By Larry Blosser

As you consider implementation of a mobile application, you'll want to evaluate not only hardware and software, but also legal issues that may arise as you move forward with your project.

Consider this checklist of legal issues as you work through your business plan:

1. CONTRACTS WITH CARRIERS

■ Wireless carriers will generally assume responsibility for obtaining any necessary FCC licenses and local zoning permits for construction and maintenance of towers and antennas, for mitigation of interference with other radio systems and for liaison with law enforcement agencies (subpoenas of billing records, wiretap or surveillance orders). Be sure these are covered in the service agreement.

■ Service agreements generally disclaim obligation to provide uninterrupted coverage or a particular level of service availability, but this may be negotiable, especially if a custom "campus" network is part of the agreement.

■ Carriers prefer to offer one-size-fits all service agreements with binding arbitration clauses, but these may be negotiable depending on the scope of your project.

■ Be sure that your contract satisfactorily addresses responsibility for costs of technology upgrades, such as when the carrier changes from analog to digital transmission, or adopts a new radio technology or billing system incompatible with the system initially delivered under contract.

- What is the carrier's contractual commitment to restore service or give service credits in the event of outage? If service is business-critical, are you given restoration priority?

- Does the agreement provide for training and for 24x7x365 maintenance and support? Is the carrier able to have 24x7x365 access to transmitters and antennas installed on your premises to maintain and repair them as needed?

2. AGREEMENTS WITH VARS AND NETWORKING CONSULTANTS

- Will there be a pre-installation evaluation of your interference environment, especially if you're operating in the 2.4 GHz unlicensed band? Unlicensed WiFi (802.11b and 802.11g) networks operating in this band are lowest in priority, after licensed FCC services, Part 18 Industrial, Scientific, and Medical devices (such as industrial microwave heaters, diathermy machines) and even incidental radiators (computer systems). WiFi networks must not interfere with any of these uses, must accept interference from them and are at risk of being compelled to shut down if interference is caused. Other bands are either less crowded (5 GHz band/802.11a) or permit only licensed (and interference protected) operation.

- Who will be responsible for obtaining zoning permits and for installation and maintenance of towers, antennas and related equipment?

- Does the agreement provide for adequate guarantees of geographic coverage, capacity and service quality, as would be defined in a service-level agreement?

- Who is responsible for compliance with environmental rules, including compliance with radio-frequency exposure limits for personnel operating and maintaining the equipment, and those individuals who may be exposed as employees, business visitors or guests?

■ Does the agreement address issues related to technology upgrades and obsolescence, whether those directly affect the vendor's equipment or software or support systems?

■ Does the agreement ensure that adequate security measures will be in place to protect business data, not only from interception over the air but also through hacking via Web interfaces or dial-up lines connected to the vendor's system?

3. GENERAL ISSUES

■ Businesses implementing mobility or wireless solutions should have in place a clearly stated and enforced policy regarding safe use, including compliance with any state or local laws concerning operation of devices while driving.

■ Be aware of any special requirements that may apply under federal, state, or local laws or regulations, and be sure they are covered in vendor agreements or company policies. These may include:

  ■ Requirements for retention of records, including instant/text messages and emails.

  ■ Detailed privacy and security obligations regarding patient, client or educational records.

  ■ Workplace surveillance and monitoring restrictions.

  ■ Privacy obligations regarding employees, including employment records.

  ■ Acceptable use policies for employer-owned devices and network services.

Larry Blosser is an attorney in the Washington, D.C., office of Gray Cary Ware & Freidenrich LLP, a law firm that represents emerging growth and technology companies. He has more than 20 years' experience in representing telecommunications users, wireless, satellite and broadband communications providers before the Federal Communications Commission and other federal agencies.

The Mobilization of Information

By Derek Kerton

We hear a lot about highly publicized 3G or WiFi networks and glamorous devices like the Palm Tungsten, BREW phones or tablet PCs. We hear much less about the unglamorous side of wireless data that is equally important: the organization and preparation of the content itself for mobile distribution.

Ironically, it is by concentrating on the unglamorous area of content organization and preparation that enterprises can differentiate themselves and have the most impact on the result. That is because the only part of the mobile solution that falls internal to the enterprise is the content. The networking is usually outsourced to telcos or wireless ISPs, and the mobile devices are selected from off-the-shelf options.

In fact, this lack of control over networks and devices actually poses a liability: How does the enterprise cope with innovative changes that it does not control? The best content mobilization solutions will be those that take into account the limitations of existing networks but are also able to take advantage of future generations of wireless networks. They will target the devices that current staff and customers use, yet also be adaptable to the devices of tomorrow.

In this section we will look at important considerations for enterprises planning to extend data to mobile users. We'll focus on Content Preparation and Content Migration since they are most often overlooked.

CONTENT PREPARATION

The real value of any mobile data solution is timely access to information. Yet information can vary wildly in quality and suitability for mobile deployment. A truly successful mobile data implementation starts with and ends with the information that is carried: how well it is prepared and how well it is presented. The network and devices that carry the information should become transparent.

No matter how fantastic the mobile device or how fast the network, poor content planning will result only in failure. In enterprises with messy data management, wireless projects hang out the enterprise's dirty laundry in public.

Most organizations store their data in "legacy systems." The term *legacy* refers to an infrastructure that made sense at some point in the organization's IT history, but now is used simply because it has the momentum of existing. Not only can the hardware become "legacy," but *data storage methods* themselves also have momentum that resists change.

For example, a company database created to store address information for U.S. customers has a field for "ZIP code," but not for "country code." Now imagine that this company begins selling to customers in foreign countries. Sales reps might enter foreign addresses by cramming the foreign postal codes and county information into the "street address" field. The result may be that mailing labels still print out in a way that appears correct, but the reality is that the system thinks they are all U.S. customers. This is dirty data. What are the risks? If the company began using a postage meter and stamping its own mail, the system would think all customers were in the U.S. and apply insufficient postage. The database could not be sorted by country or thoroughly searched. This data also would not be appropriate for wireless deployment. A sales professional traveling in Canada would not be able to use his mobile device to answer the query, "I am in postal code M4W 2R9 and have four hours before my flight. Are there any customers I could visit in the vicinity?"

The solution is to avoid dirty data every day and at every juncture. Never allow a database to be corrupted by entering information in a "kludged" manner, as the sales force above did. In this case, the IT department or the database manager should create additional fields for "country" and "postal code," or modify the "ZIP code" field to allow ZIP *or* postal codes. If solved at the appropriate juncture, the data remains clean. Once dirty, it is very expensive to manually clean a database.

For content mobilization to various device types, it pays not only to avoid dirty data but also to keep higher levels of data granularity. For example, a newspaper may store articles as word processor files. Each file is the entire story: headline, sub-head, digest (the digest is a portion of a news story, often the first one or two paragraphs, which summarizes most of the entire story to come), full story and author credits. Clean data, yes, and it may work for newspapers sharing content over a newswire, but it is not useful for mobilization to a phone handset environment. In one-file-one-story systems, the choice is to see the whole story or to see nothing. It makes more sense in a mobile environment to offer users headlines first, and, if they are interested in the headline, an option to read more. At that point, the digest could be served, and, if the reader wanted more, the full story. To offer users this choice-based delivery of the story requires that the data be more granular than one file; each portion of the content needs to be a separate entry (field) in the database.

How granular the database should be is a difficult question. The truth is that today we cannot know all the ways the data may be used in the future, all the devices that will access it and all the fields (like the "country" field case) that may need to be added. The best practice is to consult with professionals experienced in mobilizing data and using it in new ways, since they have greater experience predicting ways in which data will need to be flexible.

How data should be stored is an easier question. XML has succeeded in defining the standard for data labeling in a multi-platform, multi-usage world. XML (extensible markup language) is a way of sharing data while also sharing a description of what each piece of data is. A simple illustration is department store data regarding a box shipped from a warehouse:

This non-XML, dirty data stream is typical:

```
215525,325564125,20, 8socks 6pants 6hats
```

This XML-type data stream is a more modern way of describing the same box:

```
<record #>215525,<inventory box
#>325564125,<Total
SKUs>20,<#socks>8,<#pants>6,<#hats>6
```

In this case, the XML feed is both more granular and more identified (or "tagged," as developers say). Creating a query for a box that contains at least five pairs of socks is far easier in the XML case.

Moving back to the case of the newspaper, mobilizing a data feed that is coded in XML is far easier, since XML tags identify the subsequent data type, and we can program layout templates to include or ignore that portion of the feed. For a phone, a layout template would select only the data tagged <headline> to display just the title of a news story and ignore other tagged info. For a handheld PDA, the template would present the <headline> and the <digest> immediately, since the device is much better suited to displaying and navigating through text. For a mobile laptop (or refrigerator-mounted screen, for that matter) templates would include all portions of the XML feed including objects tagged as <image>, since there is assumed to be ample screen space and bandwidth.

The XML version of data almost always requires more storage space and more processor usage than older data storage techniques. There may be cases where the budget to mobilize data is so low that updating data structures is not appropriate, such as pilot projects, cases with few users or cases where no financial benefit is obtained from the mobile data. There are also cases where database improvements and multi-device access are not important. Generally, however, the progression of Moore's law and dropping storage prices overcome these cost barriers. Wherever data may eventually need to be viewed on multiple devices, it is almost always a better choice to store data in a highly granular manner and to tag it with XML now.

CONTENT MIGRATION

The content migration section discusses how enterprises move their information assets into the mobile space.

Where do we start? Many early attempts at mobilizing data took the shortcut of assuming the Web as a starting point. The common idea was, "If we are designing content for a small screen, let's just take the Web page we have, pull out some images and rework it into a smaller screen." This approach was intuitive but myopic. It is akin to inventing radio and deciding the radio programming should be someone reading newspapers aloud into the microphone. A new medium requires that content be created to suit the characteristics of the medium. A Web page is not a starting point, but one of many ending points. The database is the starting point.

There is additional risk when Web pages are used as content sources for wireless applications. If the HTML Web servers fail, the wireless application will also fail; however, if it were based only on the database, it would function independently. If the Web-page design is modified (as it frequently is), the wireless page will fail.

Not so with XML databases, which are modified without causing dependent applications to fail.

Using Web pages as a starting point also wastes server resources, as the Web servers are being forced to produce entire web pages (>20 kb) so that the wireless servers can pull small pieces of that data to use for the wireless pages (~1 kb). The unused data is then discarded and wasted.

Mobile applications should be built using a database as their data source.

TO AUTOMATE (TRANSCODE) OR TO COOK FROM SCRATCH?

Many companies sprung up from 1999 to 2001 offering a "magic shoehorn" that would translate Web pages onto the smaller devices. These companies offered solutions that transcode the HTML code of the Web into WML (wireless markup language) code for WAP (wireless application protocol) browsers and other code for other devices.

The promise of these companies was to release an enterprise from the complexity of presenting content on myriad devices. Most often their business models were service models, earning these enterprises the classification of wireless application service providers (WASP). For a recurring fee, such an enterprise would take your content and present it on any device. Very few pure software solutions emerged to do the transcoding job, since boxed software packages cannot keep up with the proliferation of new devices.

The quality of the WASP and other automation solutions ranged widely from quite competent to pure garbage. As this publication goes to press, most of the survivors of the WAP and dot-com blowout are competent WASPs. The solutions they provide have improved greatly in quality, now providing third generation development tools that offer drag-and-drop-type of customization capability for building a wireless application, and they seldom use a Web page as a starting point.

The decision to use an automation solution depends mostly on resource availability and quality requirements. The following table can help choose the correct strategy.

System reliability is not directly related to the decision to automate. Of course, increasing complexity by adding software or a partner will increase risk, but that risk is offset by the use of proven WASP products with dedicated support staff and (hopefully) adequate resources.

If a WASP is selected, the enterprise customer should insist on an appropriate service-level agreement (SLA), which guarantees a certain performance level and describes reparations for sub-par performance if it should occur.

Often, the best solution for an enterprise is a hybrid one where the best of both choices are combined. For example, a company like Disney could use hand coding and custom work to mobilize top brands like ESPN, but use an outsourced or automated approach for lower profile brands like Family.com. United Airlines could use a customized approach for wireless information that customers use, but use an automated wireless system for the more varied information requirements of enterprise operations.

MOBILIZE ONLY APPROPRIATE PORTIONS OF THE DATA

This simple principle requires the use of common sense. If a sales professional on the road needs access to real-time centralized data, such as inventory and current prices, wireless solutions must not require the sales professional to wade through multiple screens of undesired information.

Table 3.1 Factors in Choosing an Automation Solution

Company Requirement or Characteristic	Automation	Human design and coding
Top-quality of output		X
Low-cost solutions	X	
Expect content to be viewed on many different device types	X	
Expect content to be viewed on few devices		X
Content is the business (e.g., ESPN, the Weather Channel)		X
Content supports the business (e.g., sales-force information)	X	
Consumer oriented (e.g., airline arrival info)		X
Internally oriented (e.g., airline operations)	X	
Scarce internal resources	X	
Strange data formatting, or unusual application requirements		X
"Not built here" syndrome		X
Security or control concerns		X
Frequent design updates	X	
Short life cycle	X	
Few changes and long life		X

Unnecessary mobilized data takes up valuable screen real estate, delays delivery because of relatively slow wireless networks, and requires users to navigate around the unwanted information using the limited navigational tools of a mobile device. There is no roller-mouse, remember.

Violations of this principle are often the result of companies redeploying Web designers as mobile content page developers, assuming it's the same job. This is a mistake—they are as different as carpenter and sculptor. Although working with similar material, Web design and mobile design require radically different understandings of a user's current environment, valuations of real estate and navigation options. Some Web developers would add all the options available on the desktop version of the Web page onto the phone version. That said, Web developers are well suited to become mobile developers, given additional training and experience.

In the case of our mobile salesperson, where the required data was inventory and current price, a bad way to design the mobile information is to have him first search for the product, then receive information about that product, including images, country of origin, etc. A good way is to identify the most frequently queried data (inventory and price), and offer that on the first search results screen, with additional choices (country of origin, image, etc.) in a menu below.

CREATE REALISTIC MOBILE-USAGE SCENARIOS

Scenario planning requires developers and designers to think beyond how the information was used in a desktop environment. The designer must imagine, brainstorm, document and validate realistic mobile usage scenarios, and design accordingly. The scenario should consider the limited user interface of the mobile device.

Take the example of stock quotes. At the desktop, a common usage scenario is people looking up their stock prices to see how their entire portfolio is doing, and perhaps to trade stocks. As a result, Web pages often display stock information by showing the user's entire portfolio, the common stock indices and links for lookups and trading. That scenario is valid at the desktop, but does not apply to mobile usage patterns. A mobile user is most likely in one of two scenarios: a discussion among colleagues about a particular stock where the current price would be useful, or tracking a stock that may have moved significantly while the person was away from a desktop. In either scenario, displaying an entire portfolio is less important than viewing a single stock. Therefore, a mobile site should make it particularly easy to look up a single stock. One good design would be to offer a lookup area on the sign-in page, so users could opt to look up a single stock or to sign in for the portfolio.

NEVER FORGET THE UI

The last migration principle is to always be conscious of the limited UI (user interface) of the target devices: RIM Blackberry, Handspring Treo, Palm Vx, Nokia, Samsung phones.... No matter what the device and the claim of the manufacturer, the UI will always be far worse than the one Tom Cruise used in _Minority Report_. Don't design an application that demands heavy input, navigation or display requirements.

To refine the content for the limited UI, the best approaches are to use extensive scenario planning, with lots of "what if" questions, and to do multiple iterations of the design with improvements at each pass. Optimally, seek someone with experience designing mobile applications. This experience can be hired in house, or outsourced for the duration of the design phase.

Derek Kerton, principal of the Kerton Group, consults on business development, marketing and product strategy for wireless businesses. Derek is an expert in consumer wireless content and applications and has managed content, distribution and implementation projects for multinational wireless carriers, wireless LAN services, software companies, device manufacturers and many of the Internet's top 100 Web sites.

Transitioning Your Company To Wireless: Economic Considerations

By Joseph Sims

Wireless is, without question, the "next big thing," and its universal adoption is inevitable. By the middle of this decade, wireless communications will be an integral part of the way everyone everywhere does business. Your competitors may have started their transition to wireless already. They may even be ahead of you in that transition. The pressure is on, you're thinking, and I need to go wireless *now*.

Do you really? Why?

That's the first question that potential BearingPoint clients hear. *Why* do you need to go wireless? It's not a facetious question, and it's not meant to suggest that maybe your company *doesn't* need to go wireless. Of course it does. Why ask why, then? Because only by answering that question in the right way for your company, for your business objectives, can you leverage the technology and its capabilities in the right way—the most profitable way—for you.

"WHY" IS ONLY THE FIRST QUESTION

Finding your own answer to the question *why?* will bring the answers to these ancillary questions into focus:

- Who should my vendors be?

- What kind of internal and external support do I need?

- Where can my existing technologies, my legacy systems, fit into the picture?

- When can I expect to realize soft and hard returns on my wireless investments—not only those I will make now or in the future, but those I've already made?

■ How can I integrate this technology seamlessly into my business processes?

To appreciate the consequences of _not_ thinking about why you should make the move to wireless communications, look no further back in time than the early days of widespread Internet use. Most businesses that approached the Internet simply as the latest cool, sexy gizmo were so entranced by the hype that they ultimately missed the opportunity it presented. What made the Internet pay—what has kept Amazon.com, eBay and the other Internet success stories in business—is recognition of the more mundane (yet ultimately more pertinent) fact that the technology offered a more efficient, less costly manner of presenting information to prospective customers.

We need to think about wireless in the same terms, as another piece of the continuum. Giving people the power to access information at any hour, anywhere and in real time can promote improved response and process cycle times; greater reductions in inefficiencies, error rates, redundancies and repetitions; and increased customer satisfaction, retention and loyalty. It also can enhance decision making and productivity, by making it technologically feasible to trigger business events from remote locations. Translating that technological capability into practice depends not simply on the solution you choose, but on your company's level of centralization or decentralization, the degree of authority employees can exercise and other aspects of your business structure. Like the Internet, wireless is not an out of the box, one-size-fits-all solution or a technological magic bullet that can address non-technological issues of business systems, organization, process and procedure.

THE CASE FOR AMERICAN UNIVERSITY

In 2001, American University, in Washington, D.C., faced these questions when its campus PBX system needed extensive and expensive maintenance and upgrading. While every university used to aspire to be the most wired in the country, today the focus is on being unwired. But in academia as in the corporate world, the trick is to make sure that transition is substantive and not cosmetic. The university had to decide whether its money would be better invested in the campus-wide implementation of an enhanced wireless system capable of handling voice, data and messaging.

Further complicating the matter—and this is an issue for all universities—was that student telephone services traditionally have generated significant revenue for universities. Today, however, two-thirds of students arrive on campus equipped with their own cell phones. American University was faced with the resulting drop in revenue just at the time when needed to make a cost analysis of repairing or replacing its PBX system.

Bearing Point's challenge in tackling this ambitious project was to balance business, financial and technological concerns against the demands of a huge and diverse end-user community that included students, faculty, administrators and other university personnel. With a $2 million investment at stake and no precedent for an implementation of this magnitude in the academic world, the team began by creating a pilot program from which the campus-wide initiative could be launched.

The BearingPoint team began its work by conducting a feasibility analysis that included the business case, ROI analysis and technical/architectural assessments. After identifying the solutions best suited to meeting the university's needs and the vendors best equipped to provide services, BearingPoint launched a three-month pilot program centered around a single student dormitory and the Kogod School of Business, where the students of one graduate professor actively participated in focus group studies and other assessments of the pilot. Cell phones, wireless service and wireless access cards for use with laptop computers were tested for their ability to allow email and Internet access.

At the conclusion of the pilot program, the university had a strong and well-documented business case for extending the rollout of full wireless capability throughout the university's main campus. By September 2002, all student dormitories, the student union and many classroom buildings on the main campus were wireless-ready, and plans were in place to extend the wireless community to include the university's sports arena, off-campus sites that are integral to the campus community and its satellite facilities.

Benefits of the initiative were extensive and encompassed every area from the quality of the student experience to the reputation and financial resources of the university. The successful rollout enhanced American University's status as a thought leader in higher education, innovation, and technology—and did so while generating cost savings, creating new revenue sources and eliminating the need to continue investing in a communications system that was becoming outdated and falling into disuse.

As students move toward universal adoption of university-provided cell phones, all equipped with voicemail and a custom phone plan, American University will be able to eliminate the expense of providing landline telephone service in its residence halls. By offering rate plan discounts, it will be able to recover the revenue formerly generated by use of those landline phones. In addition, by tracking wireless Web usage and transactions, the university will be positioned to identify new revenue-generating partnerships.

The implementation also enhanced the culture and the quality of community at the school, where students, faculty and staff now maintain a point of constant connection. Students will be able to gain secure access to wirelessly enabled content on the university's internal portal, through which they can retrieve grades, view class schedules, request transcripts and use campus email. The university now has the capability to send wireless alerts to students, faculty and staff with real-time notices about class cancellations, bad weather, special events and other time-sensitive information. Even the alumni association is getting into the picture by considering ways in which the university's wireless capabilities can be used to strengthen and maintain alumni relations.

MAKING YOUR OWN CASE

What do you need a wireless communications implementation to achieve financially in order to justify the expense to yourself, your executive team, your board of directors and your shareholders?

Certainly the ability of wireless to enhance internal communications, eliminate process redundancies and repetition, increase the speed and accuracy of data and information access, and improve customer relations generates a soft return on investment. However, those returns are not the real attention-getters, particularly in more challenging economic times. Companies need to take something more substantive to the bank and to demonstrate something more significant to their executive committees, boards and shareholders. They demand that evidence of the benefits of their investment in wireless impact the income statement in terms of reduced operating expenses, achieved in tandem with indicators such as increased sales and revenue or decreases in unproductive phone time and truck rolls.

Those quantifiable metrics will justify the investment to their executive committees and board members, but they are more difficult to realize; indeed, they cannot always be realized. Moreover, achieving and quantifying those metrics can be misleading. For example, if the mobile salesperson can make one more sales call a day, what does that really mean in terms of revenue? How does that actually translate into a return on investment? Making the workforce more mobile and making information available faster are parts of the process, not solutions in their entirety. Your employees' access to information must stress not only immediacy, but also relevancy—and they must be empowered to use that information. Providing employees with faster access to information without giving them the ability to act on it will not generate a significant return on investment.

The real business case for a transition to wireless, the real return on investment, derives from using wireless to contribute to your business imperatives and objectives and to take costs out of the overall process infrastructure. At BearingPoint, we focus on achieving incremental returns on investment; our goal is to extend existing investments to include wireless, not simply to expect wireless to generate an independent return on investment.

We look beyond the technological questions to get at the heart of the business case. In the process, we may change the way you approach the procurement and support of applications and services that you deliver to your people. We examine, for example, how having the ability to execute an instantaneous real-time two-way message to every employee would impact the way you interact with your staff. Before we recommend or even assess prospective new investments, we consider, from every angle, any new ways for you to leverage your existing investments. Rather than focusing exclusively on the return on the new expenses related to your transition to wireless, we seek ways to leverage prior infrastructure, technology, process and training outlays to get an even greater return on your existing investments.

That perspective is in keeping with our belief, borne out by our experience on countless engagements, that to effect change, companies must take a holistic view of their businesses and be open to improving processes, not just latching on to emerging technologies. As with any other solution, wireless must be integrated with your business processes in order to be effective. To achieve the results our clients expect, we focus not on answering the questions we're asked, but rather on responding to the problem with which we're presented. That's a critical distinction. Extrapolating specific questions and issues that the client has and presenting those issues in terms of the overall business strategy is key.

This brings us back to our initial question: Why do you want to make the transition to wireless?

You need to explore the answer to that question in the same terms in which we would examine your company's business operations and objectives at the start of an engagement. We would look at what types of wireless capabilities that you already, directly or indirectly, support, endorse or pay for that you may not be aware of. You should work with your managers and your employees, the end users, to explore whether you're deploying your current wireless resources as effectively as possible. Do you have employees whose toolbelts feature a pager, two-way messaging device and a cell phone? You need to know what they're doing with all those devices and why you should be bearing the financial burden for supporting them all. (In fact, device proliferation is one of the areas you need to monitor very carefully as your transition to wireless unfolds.) In this manner, you can tackle the questions of what you could be doing differently, how you can consolidate, and how that consolidation would affect your business.

Wireless will become as essential to business as the Internet is. It will fundamentally change the way we do business, not only in terms of access to wireless voice or data, but also within the larger context of the whole concept of mobility. Business is moving rapidly toward the point at which operating with wireless is a necessity; one day soon, your company will not be best in breed without it. By taking the time now to approach your wireless implementation judiciously, you can identify and maximize the most compelling reasons and opportunities for mobile enablement and, in turn, implement the right wireless and mobile solutions for your company to achieve a stronger and more rapid return on investment.

HOW CAN A TRANSITION TO WIRELESS STRENGTHEN YOUR COMPANY'S PROFITABILITY?

There are many areas in which real-time, anywhere access to corporate information and applications can show benefit:

Mobile Sales-Force Automation

Your sales representatives can rapidly access customer accounts through mobile devices and get up-to-date customer information before sales calls. Contracts and forms can be stored in the mobile device and quickly uploaded to a wireless network, saving hours and reducing costs.

Mobile Supply-Chain Management

Workers can gather information about products as they move through the supply chain, from raw material procurement to finished goods. Wireless applications allow access to procurement portals to negotiate contracts for inventory or to partner extranets to react to customer demands faster. Wireless PDAs enable participation in online auctions and consideration of bids from multiple suppliers in real time, resulting in better pricing and lower transaction costs. Just-in-time supply chain information improves visibility into the pipeline, which in turn enables better decision-making.

Mobile Access to Email

Employees can stay in the loop with management, co-workers and customers when they are away from their desktops. Mobile access to email keeps them connected and enables them to act quickly on new developments triggered by email alerts. Timely access to key information allows employees to make better decisions and react quicker to customer needs.

Personal Information Management Applications

Employee productivity is enhanced by providing anytime, anywhere access to applications on the corporate networks, such as contact lists and calendar information. Extension of corporate intranets to mobile devices is being adopted quickly because it gives employees real-time access to documents, discussions and centralized information. Your organization can also realize significant savings by reducing expenditures on manuals and memos and eliminating the cost of distributing and storing standard documents and forms.

Financial Applications

Your organization can offer customers revenue- and loyalty-enhancing services such as real-time stock quotes and the ability to make mobile transactions while on the road or to check account balances at any time.

Location-Based Services

You can target a specific offering to your customers based on their actual location. In addition, your organization can extend its services to offer customers driving directions, store locations and local event information while in a mobile environment.

By answering that first question—and then those ancillary questions—for your company carefully and correctly, you'll be able to not only make the right technology choices, but also to find ways to leverage them in the most productive—and profitable—way for your unique business.

Joseph Sims is a managing director with BearingPoint, Inc. (formerly KPMG Consulting), one of the world's largest business consulting and systems integration firms. All views and opinions are his own.

Chapter 4
The Mobile Sales Team

Certainly a salesperson is one of the more obvious candidates for wireless. Salespeople are almost always on the move and every minute counts towards their potential earnings, so tools that can help them meet or beat quota are invaluable. In many cases, sales representatives don't wait for a company-wide initiative to implement wireless and mobile applications—whatever works and helps them make more sales or save more time is worth the personal cost. Many companies may find that their sales reps are already equipped with handheld devices that they've purchased themselves.

"Very often, it's the sales force that spawns the idea within an organization about how they can do business more efficiently, get higher commissions with the least possible effort," says Steve McDonald, a consultant with Optimus Solutions. McDonald says sales force automation tends to be a high percentage of the mobile application work he does.

But a company-wide mobile sales-force application has numerous benefits beyond those of individual solutions. Companies can link sales teams up to important backend data, such as inventory and order status information. Valuable sales data can be collected more easily from a device than from an employee's manual records, which makes corporate forecasting and analysis richer.

According to the research firm Yankee Group, field-sales personnel is the largest segment of mobile workers among major U.S. companies. Yankee forecasts that U.S. sales of mobile field-sales applications will grow from $132 million in 2002 to $825 million in 2006.

Salespeople are out in the field, meeting with customers, sitting in airports or cars—in other words, extremely mobile. If they're successful at what they do, then they've developed the tools that work for them and discarded the ones that don't. Sales reps and managers who are working on applications for them have to be pragmatic. Time is money, and if a mobile or wireless application is hard to use or time-consuming, then it's not going to help a salesperson achieve his or her goals. It's important to test applications that are under development with the actual users—the salespeople, in this case—who are going to integrate them into their worklife. Usability is of paramount importance, both for the sales rep and for the company that invests in the application for its sales team.

Useful sales applications can be as simple as equipping sales reps with wireless email access, and as complex as giving sales reps access to inventory shipments from China. On the simple side, for example, recruiters at Cal State Sacramento University are using a voice-enabled program from Adomo to check and respond to email. For $10,000, an organization can equip 30 to 40 users with the Adomo program, which allows field workers to listen to email on their cell phones and record responses and transmit them via sound files. Email attachments can be forwarded to any fax number.

✪ SALES GAINS A COMPETITIVE ADVANTAGE WITH MOBILE

Aside from the sheer convenience of having access to a communication device while on the go, mobile and wireless technology delivers a number of efficiency gains and competitive advantages to the sales force. While the simple ability to call a client from the road can't be dismissed, companies can now set their salespeople up with a much wider variety of wireless tools and services. These include the following:

✳ **Immediate delivery of sales leads**

Just the ability to receive email via a wireless device can be huge for a salesperson—customer questions and new customer leads can be addressed that much more quickly. For a large company with multiple sales territories, customer requests can be filtered out to salespeople in the field and dealt with more expediently than by phone. For example, queries sent to a corporate Web site address can be automatically filtered out to the wireless devices used by salespeople in the field.

✳ **Immediate inventory confirmation**

If there's any competitive advantage to be gained by being able to guarantee a shipment, wireless inventory confirmation can help. Take, for example, a company that's supplying materials to a manufacturer in times of high demand. The manufacturer might need an emergency shipment to meet increased usage, and a vendor that can provide it with on-the-spot confirmation of inventory is going to gain a competitive advantage over a vendor that must wait for a landline connection.

✳ **Immediate order placement**

A sales rep who can provide a quote or proposal and have the customer agree, then submit the order on the spot, gives immediate customer satisfaction. And in some cases, this makes money for the vendor.

Telecommunications provider Southern LINC, based in Atlanta, has equipped its 60 sales reps with a wireless order placement system that has shaved 24 to 48 hours off the time it takes to process an order. Order processing begins before the sales rep even leaves the customer site. The company's sales reps say that they're earning more respect from their clients now that they're using the wireless system—bad handwriting and incorrect delivery addresses scribbled onto an order sheet are no more. Now the reps simply type the customer's information and order into a handheld, show the handheld to the client for confirmation and submit it for processing. The company estimates it will gain $500,000 per year in faster billing activation. That's because, the sooner the order is approved, the sooner the customer starts paying for telecommunication service.

* **Increase number of customer visits**
With a wirelessly enabled handheld, salespeople can use downtime to answer emails, follow up on calls and deal with administrative tasks so that more time is freed up to actually meet with clients. More client visits means the potential for higher sales figures.

* **"One up" on the competition**
A great example of how sales reps can gain a competitive edge by employing wireless applications is the real estate market. One agent may be able to get leads on new properties and distribute them to clients faster than others with a wireless/mobile device.

* **Increased sales**
What all of this adds up to is the potential for increased sales.

Being able to confirm inventory and shipment, being able to answer a customer's questions on the spot by looking up information wirelessly (rather than waiting until she's back at the office) gives the customer a chance to say "yes" at the sales visit.

✳ **Better data analysis**

"It's simple," says AvantGo CEO Richard Owen. "Sales reps are more likely to enter correct data about a sales call five minutes after a meeting than at the end of the day." The advantage to corporate is more accurate and more timely sales data that can be used for analysis and forecasting.

✪ KEY SALES MARKETS ADOPT MOBILE SERVICES

Regardless of the industry, a company needs to look at its particular sales force and decide how mobile or wireless applications can improve the business process and the bottom line. There are some standout industries with particularly compelling needs that wireless meets quite nicely.

✳ **Pharmaceutical/medical sales**

Pharmaceutical sales reps spend the vast amount of their working hours either in doctors' waiting rooms or in transit to doctors' offices. When they do get in to see doctors, pharma reps often only have minutes to convey the benefits of a product and to answer doctors' questions. Just making use of downtime can mean a huge productivity gain for pharmaceutical companies—giving sales reps access to email and contacts data via handheld devices. But in addition to these basic functions, pharma sales reps can gain an advantage via access to more industry-specific applications, such as mobile access to a physician's prescribing history or up-to-date marketing and sales information. Having real-time, as opposed to daily, access to this information is beneficial for this industry.

Scott Hogrefe, a product manager at Viafone, which produces MobilePharma, an app for pharmaceutical sales reps, gives an idea of the value of mobile for the industry: There are 2080 working hours in a year, versus 80 hours of face time with doctors for pharmaceutical sales reps. Each hour is worth $2,000 if a sales rep is paid $160K to $200K per year. Hogrefe says it's simple to justify the cost of a wireless deployment when you look at it in these terms.

Aventis Pharmaceuticals, the maker of the popular Allegra allergy medication, has more than 4,300 sales representatives in North America. The company decided to deploy a wireless application from Wireless Knowledge that allows its salespeople to check and respond to email, access and make changes to their appointment calendars, and look up contacts stored in Outlook back at their desktops. The application helped Aventis sales reps make use of the 20 to 30 minutes of downtime in each doctor's waiting room.

✳ **Retail/wholesale**

The retail sector relies heavily on its sales teams to build and maintain distribution channels for its products. Being able to give a customer an order confirmation or a delivery date on the spot can help build a stronger relationship between a vendor and a customer.

Paveca, Venezuela's largest paper goods manufacturer and exporter, equipped its sales representatives with a wireless application from iWork Software that gives them realtime access to customer information. Sales reps can also place orders via their handheld devices. Since deploying the wireless application,

Paveca has seen a reduction of two days in its order processing time, which has increased the number of orders that leave the company's warehouse each day. Prior to implementing a wireless process, salespeople would fill out order forms while visiting customers and then they'd deliver the forms to a field office later in the day so that the orders could be manually keyed in and transmitted to the company's central warehouse. Now this process happens in real time.

* **Real estate**
Especially in geographical areas experiencing real estate booms, wireless services that allow sales agents to communicate with clients and access important services while mobile can give a competitive advantage. For example, an agent with wireless access to the real estate industry's multiple listing service (MLS) might be able to contact a new client, research appropriate properties and get that client out to see them faster than an agent who has to return to the office to do this work. Access to additional services, such as automatic loan approval, can also boost an agent's chances of getting a client into a home sooner.

* **Technology manufacturing**
In the high-technology manufacturing market, the ability to guarantee inventory and delivery is key to winning sales. A company that needs a particular product by a particular date is going to lean towards a sales rep that can guarantee the order on the spot.

* **Insurance**
Insurance sales agents are often in the field all day, visiting clients and conducting surveys of properties or cars. The ability to collect information via wireless devices and transmit it to a central processing center for policy approval gives an insurance agent

the ability to deliver accurate quotes quickly. In instances where sales agents also collect claim information, the same applies— claims are processed more quickly because the information is collected electronically and transmitted wirelessly.

CASE STUDY:
TaylorMade-Adidas Golf Delivers Premium Customer Service with Wireless System

For TaylorMade-Adidas Golf and its 110 U.S. sales reps, inventory management is a big part of the "premium golf supply" business. TaylorMade manufactures golf clubs, balls, footwear and apparel for high-tier golfers. In fact, many PGA pros use the company's products. With more than 10,000 U.S. customers and yearly sales of more than $600 million, TaylorMade is one of the top players in the golf business.

Out in the field, TaylorMade's salespeople provide valuable inventory management services to retailers without their own inventory systems (which actually adds up to a high percentage of the company's customers, many of which are smaller pro shops). In the past, TaylorMade sales reps would arrive at a retail store and spend the first one to one-and-a-half hours conducting a manual inventory of products in stock. The rep would then work with the retailer to determine the optimal inventory level and to put together the retailer's next order. Salespeople didn't have immediate access to TaylorMade's own inventory, so they found themselves making product availability promises that weren't always accurate.

The top factors that drove TaylorMade to consider a mobile solution were time savings and improved accuracy. Sure enough, when the golf supplier put mobile devices with barcode scanning capability in the hands of its salespeople, that one-and-a-half hour inventory time went down to about 20 minutes. For smaller stores, inventory times were generally reduced by about 15 percent. "For us, that's a productivity gain of 25 to 30 percent across the entire sales force," says Rob McClellan, marketing manager at Taylor-Made-Adidas.

TaylorMade sales reps can now use this extra time to look for additional distribution points or to have more valuable discussions with their existing customer base. Using ruggedized Symbol devices, the reps simply scan all the barcodes on the products in a certain store, then can analyze the data in a number of ways. Instead of computing inventory levels by hand, reps can now use mobile applications from i2 Software to quickly determine optimum inventory levels making use of auxiliary data such as the size and location of the store.

Salespeople also carry small printers that their Symbol devices can slide into to print out reports for customers. When calculating replenishment orders, reps can use their mobile devices to take into account any products that have shipped but haven't yet arrived at the store.

Time savings also come from the simple replacement of faxing in orders and having them re-keyed on the other end with transmitting orders directly from the mobile devices. Accuracy comes into play here as well, as there's less opportunity for orders to be re-keyed incorrectly.

TaylorMade started out in 2002 with a first phase that included nine users. After about three months of development, the application was deployed in a synch-only manner. Sales reps would synch their devices nightly to download updated inventory data and transmit their data on retailers' inventory and orders. As of the first quarter 2003, the entire sales staff used the devices, and they are fully wireless—with inventory and order information available in real time. "We'll always have to maintain an offline mode as a backup, though," says McClellan. "Because not all our customer sites are going to have coverage."

The wireless capability integrates a full supply and demand solution that TaylorMade is implementing on the backend, also in partnership with i2. "What the reps will have in realtime mode," says McClellan, "is essentially information on availability that's based on what's happening at the supplier in Taiwan or China."

TaylorMade considered three vendors once it had determined that a wireless solution was the best route, but choosing i2 was "something of a no-brainer," McClellan says, because i2 was already on board with TaylorMade, working on the company's backend supply and demand system. TaylorMade wanted a custom solution for its mobile workforce, and i2 was willing to cut a deal and make it a mutually beneficial project. Now i2 can market the application it developed for TaylorMade (with a few provisions that prevent i2 from selling to TaylorMade's competitors). McClellan says that none of the other vendors that TaylorMade spoke with were willing to discount the custom development.

Choosing the Symbol devices was also an easy decision, says McClellan. "A key piece of this system is barcode scanning, so we had to have good barcode scanners," he says. "We felt that Symbol makes the best barcode scanner, and we especially liked that it's integrated into the device." Ruggedization was also a key decision point: "Sometimes when you're carrying a bunch of golf clubs and other items to show a client, you might drop your device," explains McClellan. The Symbol devices can withstand a four-foot drop.

McClellan estimates that TaylorMade realized complete ROI within the limited one-year pilot phase of its deployment. And even though the company didn't consider increased sales in its ROI analysis, McClellan says that TaylorMade has seen a jump in sales since implementing its mobile/wireless system.

The feedback from TaylorMade's customers has been overwhelmingly positive, says McClellan. "Now our sales reps are going in and taking inventory quickly and essentially becoming business consultants," he says. Sales reps are now able to do a lot more analysis of inventory and make more meaningful suggestions to their customers.

"Our ultimate goal is to become so valuable to the retailer that they wouldn't consider doing business with anyone else, or if they were going to give more floor space to a retailer, we want to be first on the list," says McClellan.

THE NUTS AND BOLTS

Hardware: Symbol PPT2837 (GPRS/GSM PocketPC w/built-in laser UPC scanner)

Applications: Sales-force automation (outlet inventory gathering, accounts receivable management, order management, order track and trace, account status, order alerts)

Number of users: Over 100

Implementation time: Two phases. Phase one, 3 months; phase 2 (wireless-enablement), about 1.25 years

Cost of project: Not available

Network service provider: AT&T

VARs/vendors: i2 Technologies, JP Mobile, Symbol Technologies, AT&T

CASE STUDY:
Pepsi streamlines business process with mobile

PepsiAmericas is the second largest Pepsi anchor bottler, with operations in 11 countries, 15,000 employees, and $3.2 billion in annual revenues. Pepsi has traditionally relied on delivery agents to inform its 360,000 North American customers about products and promotions, take orders and deliver products. However, this method of operation wasn't the optimal distribution vehicle for Pepsi's wide variety of products and promotions. When the company started considering a mobile application to better manage its sales and distribution, it decided to make mobile part of a much larger business process overhaul.

Pepsi's delivery agents used to forecast their sales each day, and their trucks would be loaded with a "guesstimate" of the orders for the day. Delivery agents would head out to their 14 to 20 delivery calls each day, cut invoices based on what customers actually needed, and bring back a lot of product on the truck. There was a lot of inefficiency in this system, which required too much haulback and put excess burden on warehouse workers.

"We have over 250 SKUs, so it's very difficult for a person to manage all these products, new promotions, plus worry about delivery execution," says John Kreul, director of application development at PepsiAmericas.

Pepsi decided to transition into "pre-sell" mode—it split tasks between its sales and distribution system so that sales reps would focus on products, pricing, and distribution, and the delivery agents could focus on getting products to the customers. The company now has 1000 to 1,500 sales representatives out in the field, visiting anywhere from 15 to 25 accounts per day. The sales reps take orders on their mobile devices, then transmit them via wireless connection (or via synch mode if there isn't wireless coverage). Orders are then aggregated, routed and dispatched out to delivery drivers. Pepsi has had its pre-sell mobilized since February 2002, and the delivery side since July 2002.

"We have mobile devices at the beginning of our process, which is taking the orders, and at the end of our process, which is executing the orders," says Kreul. "And there are unique challenges with each one."

Every morning, Pepsi sales reps use their wireless devices to get current customer pricing and new product information for the day's sales calls. When the rep goes into a retail store, the handheld gives her a template based on the last six orders completed at that store. The device generates a suggested order, which speeds up the process.

"We might have 250 products, but only 10 or 15 are in that specific store," says Kreul. "So the device will pre-select all the products for that specific customer with suggested quantities."

The salesperson talks to the customer, does her "up-sells" and "cross-sells," and then enters her order into her handheld device and saves it. If she's in an area with adequate wireless coverage, she can use a wireless modem to immediately transmit her order. If not, she can wait until she has coverage or access to a landline. Pepsi's sales reps carry "portfolios" with handhelds in one sleeve, a cell phone in another pocket, a notepad to write on, and then the connection sleeve and a power cord that comes out and plugs into cigarette lighter to power both devices.

Once orders are transmitted, they go to a central order aggregation center at headquarters. Then, each mobile distribution center (Pepsi has about 90) has dispatchers that take all the orders and optimize the truck builds for the day.

On the delivery agent side, handheld mobile devices give them all their stops for the day (all the orders taken the day before by the salespeople). The delivery drivers count their trucks, adjust inventory if necessary, and make sure that the data on their handhelds matches what's in their trucks. At the end of the day, there's a checker at the gate when delivery agents get back to the warehouse. The checker inventories the truck and enters the data into the handheld, which allows Pepsi to keep better track of inventory. "It creates sort of a rolling warehouse," says Kreul. Delivery drivers can also compile their end-of-day reports on their handhelds and transmit this information wirelessly to Pepsi's central billing system.

Pepsi went with a somewhat divergent approach when it came to choosing vendors for its sales force and delivery agents applications. Kreul says they first spoke with the handful of wireless application vendors that catered to the beverage industry and narrowed their list down to two. The company's delivery software application is from Extended Technologies, while its sales application is from Thinque. Pepsi's decision to go with two different vendors for the two parts of its mobile solution was mainly based on timing—Thinque and Extended Technologies are competitors, but Thinque came out with a pre-sell application first, and Extended Technologies debuted its delivery application first.

"We have two vendors, so we can see how things go over the next couple of years and either keep them separate or consolidate," says Kreul.

Pepsi started off with a Symbol 8100 unit, but found that it was too slow and the sales reps didn't like its black and white screen. Now the company has moved on to a faster device (the Symbol 8000) that has a color screen and better usability.

Pepsi had no problem convincing its vendors to do a lengthy pilot before full implementation. Since the company's business process change was bigger than the technology change, there was no way Pepsi could go full force from day one, says Kreul. Sure enough, it took about five months to work out all the kinks in the system.

"To get user adoption and all the technology right, you really need to have that pilot stage," says Kreul. "You have to allow time for change management and for people to understand the technologies they're using."

The biggest challenge that Pepsi has faced since starting its pilot is dealing with its new business process. "Sometimes it's hard to determine if a problem is a technology issue or a business process breakdown," says Kreul. Data management has also presented a major challenge; just the process of getting all the information that sales and delivery agents need from the central system into handheld devices was huge, says Kreul. Wireless coverage was also a challenging issue; the sales pilot was in Iowa, which is a rich market for Pepsi, but a poor market in terms of wireless coverage. Agents had to turn to synch mode in many cases, so Pepsi had to make sure both methods were tuned to perfection.

"In my opinion, right now wireless should be positioned as nice to have, rather than the way to go," says Kreul.

As for the return on investment, Kreul remains tight-lipped. But he does comment about implementing the new mobile system: "It's a no-brainer."

THE NUTS AND BOLTS

Hardware: Symbol PDT 8000

Applications: Pre-sell ordering, product delivery

Number of users: 3,000

Implementation time: 18 Months

Cost of project: Not available

Network service provider: Verizon (wireless) and AT&T (WAN)

VARs/vendors: Extended Technologies, Thinque, Verizon, Symbol

CASE STUDY:
ADC takes on competitive market
with new wireless initiative

ADC Telecommunications sells network equipment, fiber optics, software and system integration services to communications service providers such as AT&T and Qwest. The company has a 250-person sales force to support its North America operations. In this industry, time is of the essence—if network equipment is failing, or a customer needs to ramp up coverage quickly, ADC needs to be able to guarantee orders and pinpoint delivery.

Until recently, salespeople did their jobs the old-fashioned way: They'd take orders and transmit them via phone, fax or landline computer. If they needed a quick answer about product availability or delivery, they'd get on the phone and try to reach someone with the information.

The idea of arming its salespeople with more precise information and saving them time on the job, combined with the decline of the telecom industry and the resultant need to improve sales, drove ADC to consider a wireless system that would give salespeople a "one-up," says Kamalesh Dwivedi, ADC's CIO.

"The telecom industry has been going through a very difficult time lately. Some call it a crash, others call it a meltdown," says Dwivedi. "It's a very competitive industry—so arming our salespeople with the ability to respond quickly in times of critical need is key."

In May 2002, ADC equipped its entire sales team with Samsung SPH-I300 devices, which integrate cell phone capability and PDA functions. The company worked with Sprint, Ubiquio, NowSpeed, Air2Web and alwaysBEthere to deploy its wireless initiative. The devices run on a Palm operating system and have 15 MB of RAM. Salespeople can use the devices to receive and reply to emails, check order status and shipping information, and access ADC's phone book of 900-plus company contacts. Users also have Web access so that they can conduct research on customers and industry issues.

"Right there, in front of the customer, the salesperson can enter the order number and within 60 seconds tell the customer where their order is," says Dwivedi. "We're supplying networks that rely on 100 percent uptime—oftentimes when we get a critical call, it's because one of those networks is running out of power and needs more equipment in a bad way, so this really gives us a leg up over our competitors."

The sales team has received the new devices and applications with open arms, says Dwivedi. "One salesperson told me it 'saved his life,'" he says.

ADC estimates that their sales team has experienced a 6 to 10 percent time savings since implementing the wireless application. And Dwivedi stresses that this estimate is very conservative: "I'd go to the bank with this number," he asserts.

THE NUTS AND BOLTS

Hardware: Samsung SPH-I300

Applications: Order status, email, Web, company phonebook

Number of users: 270

Implementation time: Two months

Cost of project: Not available

Network service provider: Sprint PCS

VARs/vendors: Ubiquio, NowSpeed, Air2Web and alwaysBEthere

Chapter 5
Mobile Hits the Field Running

When you think of mobile data users, the image that may come first to mind is the road warrior, that laptop-loving executive type with cell phone glued to ear. So it's somewhat surprising that the first wireless apps to show a true return on investment are used not by the white collars but by blue-collar folk—the maintenance engineers, repair technicians and delivery drivers.

Led by shippers such as UPS, businesses with a majority of workers out in the field use mobile and wireless applications to do everything from checking schedules to sending an automated invoice when a repair job is complete. And, while sales-force workers tend to use rather generic wireless apps—PIM, email, access to the customer database—field-force workers employ a much more diverse range of applications.

✪ FIELD FORCE: THE LEADING-EDGE MOBILE MARKET

The first wave of mobile field-force apps came way back in the 1980s, when Federal Express armed its delivery crews with custom-made handheld scanning devices that let them log deliveries, then later transfer the information to corporate systems via synch. UPS followed suit in 1993. These pricey custom devices cost the companies more than $1,000 apiece; due to economies of scale and the constantly falling price of electronics, they now cost around $500 each.

With the launch of Palm's Palm Pilot in 1996, handhelds became much more affordable. The Palm operating system was also much easier to program—businesses could easily develop Palm applications in-house, or take advantage of the many application developers and platforms, such as AvantGo. Companies like healthcare technology provider McKesson HBOC began arming delivery drivers with ruggedized Palm OS devices with integral bar scanners made by Symbol Technologies.

The advent of two-way text-enabled pagers such as Research in Motion's BlackBerry and Java-based phones that let users download and switch out applications has opened up even more possibilities. A broad swath of companies can benefit from mobilizing their field forces: Construction companies of all kinds, utility providers, commercial and home repair services, and transport and delivery services all will find substantial benefits in letting their workers send data and use applications via mobile or wireless devices.

Construction: Construction managers at residential homebuilder Amberwood Homes in Phoenix use Palm VII wireless handhelds running an application called Airwavz to update schedules and sign off on subcontractors' work. If the tile setter doesn't finish the entryway, for example, the construction supe revises a Web-based schedule, and the carpet installers get an alert telling them to put off coming to the site for an extra day.

Utility companies: Gas and electric companies employ masses of field force workers, most of them relying on paper forms on clipboards. Automating these workers using mobile devices can show substantial return on investment very quickly.

Home or commercial repair services: Letting the service technicians get quick updates, access customer information and file reports from the field saves time and can greatly reduce paperwork. When Honeywell Automation & Control Solutions outfitted the 1,400 techs who perform maintenance and repair work on commercial and industrial heating and cooling systems, it eliminated 20,000 pieces of paper each week.

Delivery services: Car rental companies have used mobile devices to speed car returns for years. Companies that make lots of deliveries have found that those devices work as well for them. Schwan's Sales Enterprises Inc., a Marshall, Minn., frozen foods distributor, provides its delivery personnel with ruggedized handheld devices and printers from Intermec Technologies Corp. The PDA lets them check customer records and take orders, while the wire-free Bluetooth-connected printer allows them to print receipts without having to go back to the truck.

✪ OPPORTUNITIES FOR SAVINGS AND EFFICIENCIES

Because field-force automation is a relatively mature industry, hard data about the return on investment (ROI) for mobile projects is a lot more readily available than in many other sectors. Many companies report positive returns within a year. For example, AvantGo, a supplier of mobile applications and infrastructure for businesses, says that its customer, McKesson HBOC, reduced imaging costs substantially by giving its delivery drivers the ability to record information and capture signatures electronically. The improved quality of the electronic records also reduced legal claims by 50 percent and reduced its delivery claims by 30 percent.

There are several reasons why mobile data services and field force workers are such a good match:

* Office time is the antithesis of the core duties of field-force workers. For technicians, installers, repair crews and delivery personnel, every minute spent at the desk is a minute when the company is not making money. For, example, research firm Aberdeen calculates that it costs a company between $30 and $60 to process a paper travel and entertainment expense report.

* Field workers are usually highly skilled—and expensive. They are often non-salaried and receive overtime pay when a job drags on. Reducing delays caused by scheduling snafus, bad information about the job or missing repair parts can make them more productive.

* Many of these workers have never been automated at all, so the boost to their productivity is even greater than for those workers who already use desktop applications.

* The information that field-force workers gather may be critical for the business's operations. Engineers at utility companies are the first line of defense against system failures; the ability to quickly report trouble spots and call for parts or repair can prevent public health or public relations disasters.

* Connecting field workers' devices to the corporate backend systems makes the data they collect available for analysis, so that the company can identify trends and project future resource needs. While it's not necessary for field workers to use mobile or wireless devices to accomplish this, most legacy field-report systems are separate from the enterprise systems; integration often takes place for the first time as part of the wireless project.

✷ When a repair crew can generate a close ticket as soon as the job is completed and send the information directly to the back office accounting application, the company can improve cash flow by collecting faster.

Because the tasks performed by field workers are so varied, from fixing sinks to filling vending machines, from watching over oil wells to delivering staples, it's no surprise that there are so many different ways to profit from wireless connectivity. Look for savings and efficiencies in the following areas:

✷ Making the field force more productive

✷ Reducing time spent on non-job specific tasks such as paperwork

✷ Reducing call center demand

✷ Generation of sales leads for upsells or replacement if repair is impossible

✷ Fleet tracking and management through GPS

✷ Reducing delivery errors and missing inventory

✷ Improved cash flow through speeding the billing process

Let's walk through how one company, Amberwood Homes, uses mobile data to cut six percent (or nearly five days) from the time it takes to build a house.

An Amberwood construction supervisor stops by the jobsite where the foundation for a three-bedroom house is about to be poured. Walking around the footings, he takes out his Palm VII, and pulls up the plan. Oops. The crew forgot that the buyers of this house requested the bay window option. This is the kind of mistake that would cost hundreds of dollars to fix after the concrete pour. The supe uses the Palm to send a fax alerting the construction crew to the error; then he reschedules the concrete pour for three days later.

The supervisor's Palm connects to the Internet and sends its information to an IBM WebSpere Application Server running the Airwavz construction scheduling application. Airwavz sends the data to an IBM DB2 database and also contacts all the subcontractors via fax, phone or email with the change. The concrete company receives the fax in plenty of time to re-jigger its own schedule. The schedule change also automatically shifts the entire remainder of the construction schedule to take into account this glitch.

Airwavz focuses on scheduling, a thorny problem whenever multiple vendors or employees must coordinate. It relies on a central database that can be accessed by a variety of devices, both mobile and stationary (fax machines, for example). Ideally, its users can receive updates in real time. But the designers realized that in the real world, not every plumber or contractor will carry an expensive wireless PDA, while even those that do will all too often not be able to get a connection. So the Airtoolz developers included the ability to receive updates via the synch method. A non-wireless user will most likely get a voice call on his cell phone from the construction supe when there's a time-critical schedule change; when he does get back to his desk to synch his PDA, he can get the entire updated schedule instead of having to sort through the scraps of paper in his pocket or the pages of his DayTimer in order to do a manual update.

It used to take Amberwood Homes about four months to build one custom home. Construction manager Dan Johnson estimates that, in the eight months since Airwavz went live, supervisors' time on the cell phone has gone down by two thirds; that fact alone makes them more productive. But there's much better news: His company has shaved a good three weeks off the home-building cycle. Yes, that's around 20 percent—with the same workforce.

The time savings comes from almost eliminating downtime waiting for subcontractors. Before Airwavz, supes and subs tried to outguess each other about when the job would be ready. The supe might tell the electricians that a house would be ready one or two days earlier than he knew it really would, because the electricians had a history of coming a day late. The electricians soon learned not to trust the supe's start date and automatically scheduled the work after the supe requested it. Now, the subs are confident that they can trust that constantly updated schedule.

Moreover, because the system creates communication logs that show when faxes or alerts went out, there's no more, "I didn't get the message." When subcontractors don't show up on schedule, Amberwood can charge them a penalty.

While scheduling is key for industries such as construction, there are many other kinds of applications that can speed the day— and bump up the bottom line—for field-force-centric companies. Automatic alerts when there's a change or important business event are another wireless application that can save time, money and frustration.

As we've seen, everyone who needs this timely information may not have a wireless device with the necessary applications loaded on it. While it's relatively easy to outfit the field force itself with the proper gear, often vendors, subcontractors, partners, or customers also need access to that info. To get the word out to everyone, you may need a universal messaging service. Although Airtoolz doesn't describe its Airwavz construction scheduling application as a universal messaging tool, it functions similarly to one when it sends faxes to subcontractors.

A true universal communications platform would automatically translate an incoming message from any source—email, voice, instant messaging application, text or Short Messaging Service

(SMS)—for transmission to the variety of devices any user might employ at that moment, such as wireless or landline phone, wireless PDA, wireless laptop, fax machine or desktop computer. To move data from one application to another, these systems frequently employ XML, a markup language that can work with many different displays and user interfaces.

Many applications that add wireless access to email also can translate and deliver email to a variety of devices and media. The ability to receive attachments that include documents, graphics, and forms adds extra value.

The City of Killeen, Texas, wanted to cut the cost of city staff's cell phone bills by reducing the minutes they used to check for messages when they were out of the office or traveling. The city used a product from Captaris that converges voice and data, letting employees, for example, receive voicemails as text messages on Internet-enabled cell phones or respond to emails using voice. The system uses the Cellular Digital Packet Data (CDPD) network. CDPD is a packet-switched network that follows Internet protocols and supports sending the same message to multiple users, a process known as multi-casting. CDPD services are cheaper than wireless voice, and providers typically charge by the megabyte of data, and allow customers to spread the bucket of data over multiple users. Captaris says the savings for using the packet network rather than the cellular voice network can run between 15 and 25 percent.

While universal messaging seems to show great promise for business communications, its development in the U.S. has been stymied by the variation in protocols used by network operators. While it's doubtful that your business application will include true universal messaging, it's smart to follow the Airtoolz example and build in support for as many different devices as you can, to make it easier for your field workers to disseminate information to their partners and customers.

Paperwork Reduction

The return on investment for mobile applications for field-force workers lies not only in giving them better access to important information, but in the simple fact that a mobile application is often the first time that their jobs have been automated at all. Many field workers juggle pencil and paper the way their predecessors did 50 years ago. Simply giving them a standard way to log information can pay off.

Because handheld devices can handle not only text but forms, spreadsheets, and even diagrams, they can do a lot to reduce paperwork. Think about all the forms, reports, time sheets and orders field workers generate. Some field workers spend as much as a third of their work time on reports and other paperwork. Any reduction in that time is an automatic increase in efficiency. That reduction can also be a morale-booster, freeing the crew from a tedious part of the job. At the same time, reports generated in the field, and either sent over the air or delivered by synching the portable device at the end of the shift, get that data more quickly into the corporate systems where it can do some good.

Providing workers with a portable, forms-based interface for entering information also greatly reduces errors. Utility companies, for example, employ technicians whose job is to roam the system, checking the status of each piece of equipment; sometimes they travel hundreds of miles in a week, using their trucks as offices. When a worker tries to scrawl notes on a clipboard that's balanced on the dashboard or wedged against a piece of machinery, it's easy to forget a decimal or write a four that looks like a nine. Later, when that tired tech has to re-key the data into the desktop computer, it's too easy to misread or guess at the figure. And of course, even if the original written entries were neat, a slip of the finger on the keyboard can create an erroneous record.

It's easy to design forms for wireless devices that limit the types of data that can be entered, force entries to stay in a certain range or disallow alphabetic entries where numerical entries are required.

Sometimes, it's important that the documents flow in both directions. Service technicians have a more pressing need than sales-people for information that can't easily be communicated by voice: diagrams, schematics, parts identification sheets. Providing them with access to manuals and tear sheets that they can easily search and call up on screen can be easier than paging through a manual and then telephoning a call center operator to place the order.

For example, First Service Networks (FSN) is a national provider of maintenance services to national retail chains such as Best Buy, Hollywood Video and Mrs. Fields Cookies, handling everything from heating and air conditioning systems and plumbing to sig-nage and asbestos removal. To provide such disparate services, FSN uses some 3,500 subcontractors. In 2000, it spent $7.5 million to install a system that links WAP phones to its Siebel CRM and finan-cial back-office applications. FSN offers the wireless application free to its subs, which are responsible for providing their own data phones and wireless Web contracts.

With this system, dispatchers don't just tell the subcontractor the customer's name, address and type of service requested. They send a history of the piece of equipment being serviced, which can help the tech identify potential trouble spots.

Information Sharing

Mockler Beverage Company, the exclusive Anheuser-Busch dis-tributor for Baton Rouge, Louis., and the eight surrounding parishes, used to employ the "peddle" method for sales: Drivers would make an educated guess about what stores on their route needed, then load their trucks accordingly. "Loadbacks," inventory that was loaded on the truck but came back unsold at night, reached as high as 50 percent. In 1998, Mockler automated its sales and delivery processes, providing its salespeople with Intermec ruggedized handheld computers with integral scanners and software designed to automate merchandising, sales and deliveries. Now, as drivers call on their retail clients, they log orders on the handheld and then send them to the warehouse by attaching a Nextel wireless phone to the handheld with a cable.

The data goes to the central computer at the warehouse, which builds the next day's delivery loads. That same information goes to Mockler's accounting, management, pre-sales division and delivery drivers. This system lets the warehouse start building the orders the same day they arrive, which in turn lets the drivers get out on their routes earlier the next morning.

In the warehouse, workers use the same devices with scanners to track inventory in storage and on the trucks. The company estimates that the devices save one staff hour a week of inventory time.

Reducing Travel Time

Reducing travel time is another important way to keep those technicians busy making money for the company, and this is another place where wireless applications have proven themselves. Global positioning systems (GPS) let dispatchers quickly find the field worker who is best placed to make a service call. (For more about keeping track of trucks and other vehicles, see Chapter 9.) Combine this GPS system with the ability to send roving technicians all the information they'll need to handle that call superbly, and you've created quite a customer service edge.

Automating Time Sheets

Complaints about tardy appearances by repair personnel or over-statements of hourly charges can be minimized by automating the reporting process. When field techs have to track their hours by hand, they may scrawl illegible notes, intentionally or unintentionally misstate times, or put off the task until the end of the week and try to re-create their schedules from memory. This behavior puts the company at a disadvantage when a customer disputes the bill. Applications that wirelessly send service requests to technicians also automatically log the calls. When technicians can report their arrival and job completion times by tapping a form on a screen, they'll do it. Advanced—and expensive systems—that include a GPS system on the service truck can provide full reporting of service visits with no need for the tech to do anything.

Providing Maps and Diagrams

Carrying enough maps and schematics for every eventuality is close to impossible for field-force workers. And even when techs have them, they can lose precious time flipping through them to find the right page. Wireless access to searchable charts, maps, and schematics can pay for itself, especially when the cost of not having the right information is high. For example, Northeast Utilities, which serves 1.7 million customers in New England, worked with iAnywhere Solutions to develop Direct Burial Map Locator, a laptop-based wireless application that lets techs call up maps that pinpoint locations of NU lines and equipment.

⊙ INDIRECT BENEFITS OF MOBILE APPS

Reducting call center costs: Wireless data services can also reduce call center costs. Sears, Roebuck's HomeCentral repair service began using a wireless system from vendor Ubiquio back in the early 1990s.

The system lets the repair technicians look up parts on wirelessly enabled laptops and then automatically order them. Before it went wireless, Sears repair techs made 1.4 million calls a year to the call center—at a cost of between $2 and $4 per call. The call center operator would take down information, look up parts, consult with the tech and then order them. Letting technicians access the parts database, choose products and place the orders wirelessly cut such calls by 80 percent.

Improved cash flow: Repudiation of charges is a huge problem for companies that make deliveries, because changing schedules and staff turnover can make it hard to find out who received merchandise. Many custom handheld devices for field-force workers come equipped with a micro-printer that lets them hand the customer a bill or statement when work is complete, then capture the customer's signature electronically. Having the sign-off information and signature available in the company's accounting and billing system reduces repudiated bills, which can otherwise take weeks to resolve.

At the same time, when mobile devices automatically send information to the company billing system, it can save as much as ten days in the billing cycle, getting the bills out faster so the money can come in sooner.

More sales leads and upsells: When a technician finds a machine that's beyond repair, or on its last legs, she can send an alert to the sales division, which can quickly give a quote on new equipment, minimizing downtime for the customer and the possibility that that customer might look for a different supplier.

Improved visibility into the enterprise: A well-designed field service application should be able to connect with other company divisions. For example, a smart sales manager might look at the service requests for a key corporate customer, and determine that the customer would save money by purchasing new equipment that was under warranty. While some of this information might be already available, often this kind of key information never makes it into the company's databases.

Increased customer satisfaction: Getting data from the field force into the enterprise systems can provide strong customer-service benefits when the company provides a Web-based tool for self-service tracking. For example, in March 2002, Boise Office Solutions of Itasca, Ill., a multinational distributor of office and technology products, gave its 1,100 delivery drivers stationed at 38 different distribution centers mobile computers and bar scanners from Intermec Technologies. Boise's drivers use the scanners to collect proof-of-delivery information. That data will also be fed into Boise's order-processing system, so customers can track the status of their office-supply deliveries.

✪ THE BAD NEWS

Despite all the opportunities for efficiencies and cost savings, outfitting your field force with wireless devices and applications is not a slam-dunk. There are as many barriers to success within this sector as there in for any other: lack of coverage, bad connections, dropped calls, expensive devices, clumsy software. The single biggest issue, as one executive puts it, is "Coverage, coverage, coverage."

Field-force applications face some unique barriers, as well. The first is resistance from the workers themselves. Some field workers may not be especially computer literate; they may be unwilling, or at least uneasy, about having to learn the system. In order to use wireless applications—or any other sort of computerized or automated system, for that matter—users must conform to the system and may have to change their work procedures and habits. This is especially true of the more tradition-bound industries, such as construction. Workers may protest, "But this is the way I've always done things." A couple of Amberwood Homes' construction supervisors chose to move on rather than use its mobile scheduling system.

Field workers are used to working independently without supervision and may resent what they see as "being told how to do my job." Until they get comfortable with the system, it may seem like added work rather than a time-saver. When the system is connected to other e-business applications, managers can use it to plot individual performance against the norm, which can make workers feel like Big Brother is watching.

Getting partners to use the system can be even harder, especially if they'll be required to spend their own money on devices or service. For example, First Service Networks, the retail-store maintenance provider, found that despite the application's utility, there were four reasons why its subcontractors didn't take advantage of WAP phone access to its CRM application:

* A few of FSN's subcontractors didn't have good enough basic cell phone coverage to make sending data feasible.

* Some subcontractors had existing wireless phone contracts with carriers that didn't support WAP.

* Some used carriers that supported WAP, but hadn't provided their employees with data-enabled phones, and didn't want to spend the money to upgrade.

* Some used data-ready phones but couldn't make a business case for paying the premium charged by their carriers for wireless Web access.

To increase employee acceptance, managers should recruit influential field-force workers early in the design process. They'll provide valuable information about how wireless applications can benefit them best, and, if they buy in, they'll talk up the benefits to their co-workers. Be very open about the wireless project and the expected savings and benefits, and stress that they will benefit not only the company's bottom line but also the employee's day-to-day work. Benchmark to make sure that, after roll-out, workers truly are more productive. Identify those who aren't seeing benefits and offer them additional training.

A workforce that has never been automated or computerized may have a steeper learning curve than traditional desk workers. But taking care to foster adoption of mobile devices will lead not only to excellent productivity gains, but also to better morale when workers realize their paperwork has vanished.

CASE STUDY:
Wireless Makes Fast-Food Techs Faster

ParTech Inc. provides point-of-sale (POS) systems for fast-food retailers, including Taco Bell and McDonalds. Without the ability to log orders and ring up the charges, the business crashes to a halt, so technical support and customer service for the POS are critical. ParTech employs some 130 field engineers in the U.S. to service customers, each one responsible for an area with a radius of about 200 miles from the base.

When they have a problem such as a register locking up, customers first work with technical support staff at the call center. These are highly trained staffers who use their own experience and a computerized "knowledge base" to try to get the customer's POS machine back online. If that fails, they generate a work order and assign a field engineer. ParTech uses Clarify ClearSupport and Clear-Logistics to record call-center and field-service information. However, the field engineers had no direct access to the system.

"The way we handled communications created a lot of redundancy," says Bill Hauck, manager of business applications for ParTech. When the call center engineer determined that a service call was in order, he would create a computerized work order using Clarify; the application would show information about the customer including any SLA. The call-center engineer would then call the voicemail of the field engineer assigned to the job and read the information from the screen. Leaving the voicemail would trigger a numeric message to the field engineer's pager, who would know to call in and listen to his voice message. If he needed more info, he would telephone the call center. If, on the jobsite, he found he needed parts to complete the work, he would again call in. A final call was necessary to report that the repair was completed. The call-center staffer would then enter details into Clarify. One customer call would typically generate four additional calls to the call center.

Most customers have SLAs with the company, some with as little as a two-hour response time at any time of the day or night. As soon as they call customer support, the meter starts ticking. It took ParTech an average of from one to two hours to retrieve the necessary information about the customer; calls to check parts availability, parts orders and closing the case added another four hours. There were occasions when ParTech failed to meet its SLA. It seemed that the field engineer sometimes spent more time communicating with the call center and getting information than in the actual service.

In September 2002, ParTech rolled out a wireless communications system from Xora. The system lets field engineers communicate directly with the Clarify system using WAP phones. Now, when a call-center engineer creates a work order, the field engineer gets an automated alert that includes details of the service request. This message is "actionable"; it can contain clickable links to more information or applications. If he accepts the order, he enters the estimated time of arrival, which is saved into the system. If he rejects the order, or doesn't respond in the prescribed time, the system automatically assigns the job to another engineer.

Call-center engineers can now spend more time helping customers and less time taking messages from the field engineers. "Once the call-center agent dispatches it," Hauck says, "that's the last they see of it."

Meanwhile, the customer receives an automated voice alert with the estimated time of arrival. The field engineer can update the ETA if necessary, and the customer is automatically kept informed. He can also query Clarify to find out more details about the customer.

Once on site, the engineer can use the WAP application to log what parts were installed or to order necessary parts. Clarify automatically notifies the warehouse to restock the parts on the engineer's truck. When the engineer is done, he informs the system, and it logs the repair as complete, noting the time.

Implementation took four months, because ParTech asked for modifications to Xora's application, based on feedback from the engineers. ParTech needed the ability for engineers to create emergency parts orders through Clarify, rather than having to call the call center. This is faster and also limits ordering errors. Hauck doesn't have metrics on how often wrong parts are ordered, but, he says, "Even a small percentage of errors on emergency parts hurts." ParTech also asked Xora to let managers access the system from their WAP phones, so that they could check open work orders and redirect jobs even while they were out on customer calls.

Hauck chose to mount the wireless access to Clarify on WAP phones for a couple reasons. For one, the field engineers already carried cell phones, so he wouldn't be loading them with another device. "If we added another device, they'd look like Joe Geek with the tool belt," he says. The advantage of using WAP for wireless access, he says, is that it's device independent. "The beauty of it is that as the [available] hardware changes, we can change also. The application will follow right along."

True wireless connectivity lets his field workers get information in real time, and Hauck says it's also increased customer satisfaction because they, too, can log onto the Web to follow the status of their service requests. "When we won some business from a competitor," Hauck says, "one of the complaints they had was that they couldn't get real-time information because our competitors' field engineers used the synching process at night."

Hauck expects to pay off his investment—which he'll only say was "reasonable"—in 18 months. Payback will come from increased productivity for his field engineers, better inventory control and call-center efficiencies. Here's where the savings will come from:

✳ Information retrieval and database interaction times were reduced from 6 hours to 20 minutes.

✳ Field engineers became more productive and could handle more cases each week.

✳ Call-center staff who handled phone calls from field engineers were re-deployed to other areas that would increase customer satisfaction.

✳ Customer-satisfaction levels increased because of faster response and improved customer communication.

THE NUTS AND BOLTS

Hardware: Nextel 55SR phone

Applications: Clarify (other applications to follow)

Number of users: 140

Implementation time: Two months for pilot; additional four months for phased nationwide production rollout

Cost of project: $200,000

Network service provider: Nextel

VARs/application vendors: Xora

CASE STUDY:
Mobile Forms Keep Water Company Up to Speed

Southern California Water Company, the second-largest in California, provides water to 27 different water systems in California and Arizona. The water production facilities are groundwater wells; each site consists of a fenced area with a large well, a pump, electrical panels and piping. The station pumps water out of the ground, disinfects it, and pushes it into the pipes that send it along.

The SCWC's frontline workers, "pumpers," travel from pumping station to pumping station reading the meters. They used a piece of paper with the site's name to record how much water had come through, what the water level was, how many hours the pump ran, and how much power it had used, collecting anywhere from 30 to 85 data points each day.

After they'd looked at all their sites, they'd go back to the office and manually input all those readings into the computer. If one of the readings didn't look right—for example, if production at a particular pump had varied wildly from the day before—they'd have to go back out to the field to investigate. All in all, each pumper spent about an hour a day filling out forms. Meanwhile, SCWC had to subcontract out much of its maintenance work.

Toward the end of 2000, water supply superintendent Paul Schubert decided that that hour a day could be better spent. "Collecting reads is one of the most important things they do," Schubert says, "but there are other thing they could be spending their time on—plant maintenance, preventative maintenance." He decided to automate the data-collection process.

SCWC already used its Supply Quality Utility Informational Database (SQUID), a custom SQL database developed in-house in 1998 that kept track of all the information taken from each meter read: hours the pump ran, kilowatts used, volume pumped, and static water levels. Schubert worked with applications developer DST Controls to develop a custom application for the pumpers, running on Symbol handheld devices with integral bar code scanners with the Palm operating system.

The first system, Arden Cordova, went live in April 2001. With the new system, pumpers simply point the scanner at each pump's barcode (Schubert himself outfitted each pump with a bar code, printing the files sent by DST on his office printer, then inserting them in plastic luggage tags) and the device brings up the right form onto the screen. The pumper simply taps the readings into the form. At the end of the day, data is synched to the SQUID database. (For an example of how this process could be further automated by including wireless connections to sensors on the pumps, see Chapter 7.)

The pumpers saved more than that hour a day they used to spend on paperwork. They can log their reads quicker, because tapping the screen is faster than writing numbers out. The barcodes "reduce the typo factor," Schubert says. Each form is preset with ranges, and if the pumper tries to input something outside the range, it prompts him to double-check. Pumpers can immediately compare the new read with the previous one, and flag problems on the spot. When they synch at day's end, the system automatically generates a work order for the trouble spot. "That prevents a problem from continuing for a longer time," Schubert says.

The information can also be more quickly shared with other SCWC divisions. Engineers, supplies superintendents and the financial department get current information every day, where previously they might wait up to 15 days. This lets them see trends in usage so that they can be prepared to meet peak day and hour demand while keeping reserves for emergencies such as fires.

The whole system cost less than $20,000. SCWC saved some money by making use of the SQL database's open architecture, which didn't require any license fees for the new app. Now, he says, "I could roll this out to any water system in the state. All it takes is $650 to buy the device. I have a CD with the application; they just download it and off it goes." Unfortunately, Schubert had to wait to roll out the mobile application to other stations, due to limited company resources. He expects payback on the existing deployment in a year and a half to two years, through the savings generated by having the pumpers use the time they've saved to handle maintenance in-house.

THE NUTS AND BOLTS

Hardware: Symbol handhelds with integral barcode scanners

Applications: Palm OS, custom application

Number of users: Two

Implementation time: 60 days for software development; 2 weeks to deploy

Cost of project: $20,000

Network service provider: None

VARs/application vendors: DST Controls

Chapter 6

Wireless Local Area Networks

Wireless local area networks (wireless LANs, or WLANs) are one of the hottest wireless technologies around right now. They're turning everything from sports stadiums to coffee shops to corporate campuses into wireless access points, also known as "hot spots." There's obviously a different draw for all of these venues—in some cases, it's a customer-retention feature, while in others it's an issue of employee productivity.

WLANs are poised for double-digit growth, according to Gartner, with healthy growth expected through 2007. "The initial strong growth in the WLAN equipment market is being driven by the mobile data-connectivity needs of professional portable PC users," says Andy Rolfe, principal analyst for Gartner's worldwide telecommunications and networking group. "The increase in WLAN-enabled mobile PCs and PDAs will drive demand for WLAN access in a variety of locations to support mobile access to business applications."

Gartner forecasts the penetration rate of WLAN into the professional mobile PC installed base will grow from 9 percent in 2000 to almost 50 percent by the end of 2003, and it is expected to surpass 90 percent by 2007. "As WLAN equipment prices continue to fall and speed increases, wireless solutions will become a viable alternative to wired LANs in small premises," says Rolfe. "This is because bandwidth demands are lower in small sites, and the cost of cabling for wired Ethernet is higher than in larger premises."

Most of the WLANs out there today are based on the 802.11b standard, sometimes called WiFi, which operates in the unlicensed 2.4 GHz spectrum, the same frequency band used by cordless phones, microwave ovens, and Bluetooth. A WLAN may use one or more "nodes," wireless transceivers that relay signals to and from Internet servers. Each one has a limited range, about 300 feet. While a string of nodes could theoretically cover a large area, WLAN technology typically is used to provide coverage for discreet properties, such as an office park, college campus or even a single store.

A WLAN works much like a wired network: Devices equipped with the functionality to operate on the network can plug in, transmit data, and access network services such as databases and email. Instead of using phone lines or cables to transmit data, a WLAN uses radio or infrared signals. These WLANs offer broadband capacity, at speeds up to 11 megabits per second, as fast as the new networks U.S. telcos hope to roll out by 2004. There are also a number of standards in development that may usurp 802.11b. 802.11g is the forerunner in this group, recognized as the "next generation" of WiFi. It's compatible with 802.11b, and uses the same spectrum, but offers data rates up to about 54 Mbps. The 802.11g standard also addresses some security issues inherent to 802.11b.

Research group In-Stat/MDR estimates that the number of public-access WLAN locations will grow worldwide from 2,000 in 2001 to 42,000 in 2006. Service revenues should see a similar spike—from $11 million in 2001 to $642 million in 2006. Though 2002 was billed as "the year of the WLAN," the industry is still young and under development. There are standard and security updates yet to come that will advance the technology and eradicate many of today's concerns about security and performance.

Nevertheless, many companies are adopting WLANs right now. Some are giving it a go on an experimental basis, trying just a few applications with a smaller set of employees. And given today's economic climate, many vendors are more open to offering these types of extended product trials. Other companies and organizations have gone through the evaluation process and decided that WLAN technology and applications do offer them a significant advantage in the marketplace and gone ahead with full-scale deployments.

A knowledgeable vendor or consultant will be able to advise a company about whether it's a better idea to wait for an updated technology or to move ahead with what's available today. Many applications based on today's popular 802.11b WLAN standard are equipped to evolve with new standard iterations, so if a company can see a clear business case for deploying a WLAN, decision makers don't have to worry about scalability.

✪ KEY BENEFITS OF WLANS

WLANs are being implemented in scores of business settings; in some cases, the expected return on investment is as simple as improving business meetings by giving workers easy access to network data on their laptops. But the potential benefits of WLANs reach far beyond simple wireless computing.

Quick Access to Data

For many employees, just the ability to get information in real time is key. Whether it's a salesperson at the airport who can close a deal because he can check inventory, or a worker in a car manufacturing plant who can check the availability of a certain part without leaving his work area, quick access is often the most important reason to deploy a WLAN or set up workers with WLAN-enabled hardware.

General Internet/Email Services

In retail settings such as coffee shops or even airports, simple Internet access is a top attraction. To take it a step further, setting up workers with access to corporate email can prove invaluable if they're in a position to make use of interchanges on the spot (sales representatives, for example).

Wireless Ordering

While definitely one of the more marginal uses of WLAN right now, wireless ordering has been deployed in some sports stadiums. Fans can order their hot dogs and beverages from wireless terminals at their seats, saving them from missing game highlights and saving wait staff from navigating through the crowded stands to take orders.

✪ THE MARKETS FOR WLANS

Just as with any other wireless technology, WLANs have distinct benefits for various market segments. The technology is also very popular in the residential market. Some of the more traditional WLAN markets include factories, warehouses and distribution centers.

Hospitals

The primary drivers for hospitals to deploy WLAN-based applications are better record keeping, better patient care and more time for doctors. In the Yankee Group's 2002 annual survey, 47 percent of the healthcare institutions that the research group spoke with

had already deployed a WLAN. Younger doctors especially are recognizing the benefits of being able to access medical records and
data in real time. Many of them are arriving on the job with their
own handheld devices, so when their employers implement an
organization-wide application, the adoption curve isn't too steep.

The healthcare environment in general has adopted wireless
technology faster than other markets, because its needs are so immediate. The Health Insurance Portability and Accountability Act of
1996 (HIPAA) also set forth a requirement for healthcare organizations to use electronic documentation over paper-based practices. However, healthcare also faces higher security risks than other
sectors; there's federal regulation involved and serious privacy issues.

In one respect, the higher standards of the healthcare market
are helping the industry as a whole. On the way to reaching healthcare's high security and data transmission standards, other industries' needs are met. There are numerous vendors who are innovating
for the healthcare industry and then serving high-quality products
to auxiliary markets as well.

Public Services

In the public service arena, it's all about saving people time. Whether
for a city-wide service that benefits citizens, or a few key access
points that make employees' jobs easier, government and city agencies are seeing the benefits of WLANs. The city of San Mateo has
implemented a system of WLAN access points for its police officers to use while in the field. If an officer needs access to the police
station's internal network or needs to download photos, she can
now stop at one of the access points and user her laptop computer
to do this work, instead of driving back to the station. Officers can
even write reports while out in the field and then stop at one of the
access points to file and print the reports remotely at the police

station. San Mateo received a federal grant from the Office of Community Policing to fund the implementation. While security was a big concern with this project, Lt. Wayne Hoss of the San Mateo Police Department says that his team was able to implement a combination of off-the-shelf security programs that they feel are sufficient to protect sensitive police data.

The city of San Diego has launched a drive to offer free, wireless Internet access throughout its downtown area. San Diego started with three locations near popular downtown buildings—within 1,000 feet of each of these "hot spots" wireless device users can tap into an 802.11b network.

Passengers in Switzerland's railway stations can access the Internet via their 802.11-equipped laptops. To access the network, provided by Swiss wireless carrier Swisscom, users submit their mobile phone numbers to get a code that grants them access for a single session. The WLAN service is also available in Swiss airports, hotels and conference centers.

Universities

Universities are using WLANs to give students access to numerous sets of data and to give visitors information as they explore the campus. For example, American University in Washington, D.C., deployed an 802.11b network to give its students and faculty email and intranet/Internet access via their laptops and other mobile devices. Students can receive class schedules, course availability information, grades and transcripts via their wireless devices or laptop computers. American University also uses the network to send students alerts about class cancellations, bad weather and event information. Other universities are using their 802.11 networks to distribute time-sensitive coupons to students for on-campus restaurants.

Harvard Medical School set up an 802.11b network and deployed a set of applications from AvantGo that give its busy medical students access to class schedules, hospital case log notes, lecture notes, animated anatomy illustrations, course evaluations, exam calendars and last-minute announcements. Harvard also used the wireless network to replace its paper-based survey system. It was this aspect of the deployment that brought real ROI for the university; in just six months the school realized a 50 percent savings from replacing its paper surveys with an automated, mobile solution.

Airports

Where else but an airport does a business traveler have so much free time? With flight delays, layovers and new security restrictions, an airport could be a goldmine for a service provider offering just the right functionality. It was the airlines who realized this first, and started rolling out WLAN access in their terminals and lounges. But a bit of contention began to develop between airport authorities and airlines, once the airports realized how chaotic it could get if each airline had its own wireless network offering. Not to mention the cut of revenues the airports were missing out on by not getting into the action. The Wireless Airports Association (WAA) is pushing for a standard for WLAN implementations at airports, which would make things easier on travelers. But a few airports nationwide have gone ahead with implementations in the meantime.

There's also the possibility of going beyond the basic WLAN access for business traveler services: Providers could broaden WLAN services to include specialized applications for the airlines, the airport concessionaires, airport employees and security personnel.

For example, at Minneapolis-St. Paul International Airport, travelers, airport concessionaires and airlines have access to a wireless network hosted by Concourse Communications and iPass.

Travelers who have a laptop or a handheld device equipped with a WiFi network card can obtain one-day access to the network for $7.95, or they can choose from various subscription plans. Concourse is working with airlines and airport concessionaires to develop useful business and security applications that run on the wireless network.

Hotels and Conference Centers

Again, hotels and conference centers have a clear business case in mind when rolling out WLAN services: the business traveler with time on her hands. Many hotels are now offering wireless Internet access within rooms and often all over the hotel grounds.

At the Hilton Antwerp in Waterloo, Belgium, guests can access a WLAN from anywhere within the hotel grounds. The hotel manages a portal that provides guests with information on hotel services—updates on a guest's mini-bar account, for example. Guests can also choose from various options for Internet access, depending on how many days they're staying at the hotel.

The Royal Sonesta Hotel in Boston is taking a different approach with its WLAN. The hotel offers wireless location-enabled guided walking tours to patrons and visitors at its modern art collection, which is on display throughout the Royal Sonesta's public spaces and restaurants. Visitors can pick up Compaq iPaq handhelds configured with 802.11 wireless cards, and read about the artwork they're seeing as they browse the exhibits on the first three floors of the hotel. The payback for the hotel is that they've seen more visitors eating in their restaurants since the deployment of their WLAN-based tour service.

Corporate Campuses

In the day-to-day corporate environment, a WLAN often simply takes the place of a landline network. For a company with a large campus and multiple meeting rooms, it allows workers to access files and databases while away from their desks. For IT workers, the ability to access information and deal with network issues while on the go is invaluable.

Retail

The retail industry has deployed WLAN services and applications both to improve its own in-store business practices and to lure in customers who want free wireless Internet access. Starbucks is one of the most well-known examples, with WLAN services available in 1,200 stores, though plenty of independent coffee shops and cafés have followed suit with free wireless Internet access for their customers. Schlotsky's Deli has launched WiFi services in many of its restaurants, for those who want a "working lunch." And in downtown Palo Alto, there are so many retail locations offering WLAN access that you can get a signal almost anywhere in the downtown area.

✪ THE SECURITY QUESTION

Just as with any emerging technology, WLAN doesn't meet the security standards of many discerning organizations. Security improves with every iteration of the standard, but the popular 802.11b version has been lambasted in the press for its lack of security. Often, the horror stories relate to organizations that unpacked a WLAN, installed it, and never even turned on the most basic security features that were included with the product.

Even aside from making sure that embedded security features are active, there's still a lot a company can do to ensure WLAN security. First, organizations need to decide how much security is enough. It all really depends on what types of applications are going to run on the WLAN. For example, most hackers don't care about banal company data. So a company that's using its WLAN for scheduling updates may just need to turn on wired equivalent privacy (WEP) security. Of course, if an organization is dealing with really important data, such as financials or credit card information, then coming up to speed on the strictest security products available is imperative.

Proper network management is a big part of addressing security; if not managed properly, WLANs can suffer from performance issues and are prone to intrusion. The National Institute of Standards and Technology recommends that organizations implementing WLANs take the following precautions to protect their networks.

* Understand the architecture of your wireless network.

* Maintain an inventory of the wireless devices you have out in the field.

* Back up data frequently.

* Perform regular security testing and assessment of the network.

* Upgrade security and apply patches to discovered vulnerabilities when necessary.

* Keep pace with the industry, so that your network can stay up to date.

The best place to turn for answers to security questions is a wireless consultant who deals with these issues on a day-to-day basis. Though within a year, the market for WLAN security should whittle down to a few proven, practical security protocols. Companies such as Computer Associates are also working on "predictive capabilities," which would provide intrusion detection on a WLAN.

Avoiding the Tootsie Pop Syndrome

By Christopher W. Day, CISSP

"Can you believe this wireless network? In less than a minute, I'm on the Web. Look, here are their servers ... I can see the HR system—here's the employment contracts folder ... and this looks like the customer database ... there's the mail server ... I wonder if I can log in to it...." –Conversation with a hacker in a parking garage (August 2002)

While wireless networking has advantages and strong appeal for people who want access to network resources and the Internet without being physically connected to a location, it also has a darker side when it comes to security. The benefits of wireless networks— access through walls, portability, plug-'n'-play setup, easy network access, wide footprint for users to connect—also make them highly vulnerable to unauthorized eavesdropping.

In May of 2002, an anonymous user posted on the Security Focus Web site that he detected credit card numbers being transmitted via an unencrypted wireless network at his local Best Buy. Best Buy had implemented a wireless network to support a wireless cash register system but had failed to implement basic security on this equipment. Other Security Focus readers quickly determined that many Best Buy stores throughout the country had the same vulnerability.

The situation was exposed by the national media which led to deactivation of the equipment and a series of press releases from Best Buy attempting to explain the problem. It is not known if any credit card numbers or other information was stolen.

Wireless networks use radio waves to carry information. Radio waves pass through solid objects such as floors, windows and walls. This is in contrast to standard wired networks that use a physical wire terminating into fixed ports. In this article we refer specifically to the 802.11 family of wireless products, often referred to as "WiFi." While some of the technical specifications will differ, many of the principles and strategies discussed here apply for networks using other wireless protocols.

MISCONCEPTION #1:
"Why would anybody want to break into my network? We are only a <insert your industry> company; we don't do anything that would interest a hacker...."

For companies that are not in a highly visible industry—such as entertainment, banking or government—this may seem a reasonable question. However, my experience securing networks for hundreds of clients throughout the world has demonstrated that many intrusions resulted from an attacker who took advantage of a "target of opportunity." In other words, intruders are not necessarily targeting specific companies to attack, but instead are searching for vulnerable networks.

We've found that most opportunistic wireless attackers use automated tools and fairly simple techniques to scan large areas for networks with vulnerabilities. Only after an exploitable network is discovered does the hacker then target a specific company's network.

Asgard Group was hired by a large national retailer to perform a security assessment of its network. During a site visit to one of its stores, we realized that the company was using a wireless network to support mobile cash registers and price scanners. Unfortunately, the wireless network was connected directly into the internal network and no security at all had been implemented on the wireless devices.

Our engineers (as well as any intruder) were able to easily access the network from the street and parking garage. Once on the network, we had free access to surf the Internet and peruse the retailer's financial and human resources systems.

This wireless network had been in place for almost a year. Our security team determined that intruders had compromised multiple servers, most likely via the wireless network. An extensive cleanup and restoration project was then necessary to secure the network and restore the compromised systems.

The attacker isn't always after you. He may be after your processing power, bandwidth and storage resources. Goals for hackers include:

- Innocuous theft of services such as free *web* surfing over your internet connection.
- Mischief and disruption of your network.

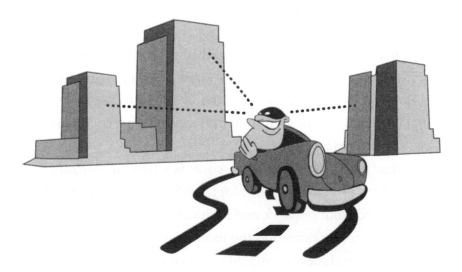

War driving is a popular activity among hackers seeking vulnerable wireless networks. War drivers use a laptop, an inexpensive wireless network card and free software to "sniff" for open networks. Once located, malicious war drivers then exploit these networks for pleasure or profit.

- Questionable entrepreneurial e-commerce, like using your servers to host pay-per-view pornography sites, sell stolen DVDs and credit cards

- Outright exploitation by attacking your systems and/or using your network to attack other companies

MISCONCEPTION #2:
"We're protected. We have a firewall and/or anti-virus software installed."

Many companies have created a network security structure that is like a Tootsie Pop, with a hard outer shell (the firewall) protecting the soft, chewy inside (the servers and data). Often, both technical and non-technical employees mistakenly believe that installing a firewall or anti-virus software provides adequate security.

Firewalls and virus protection are just part of the security process. Prevention, detection and response are all needed to make networks more like *jawbreakers*—hard all the way through.

Prevention keeps unauthorized users from accessing systems. It is analogous to locks on doors that are used to keep intruders outside of your offices. Common prevention technologies include firewalls and passwords.

Detection is the alarm that sounds when an intruder has bypassed the prevention technologies. It's the equivalent of a burglar alarm for your network and includes intrusion detection systems, anti-virus software and log monitoring.

Response includes automated response to intrusions and manual intervention by skilled security experts.

For security to be effective, detection and response includes periodic vulnerability assessments by experts, ongoing monitoring for attacks, and structured response mechanisms.

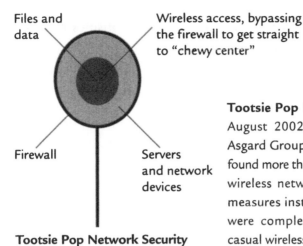

Files and data

Wireless access, bypassing the firewall to get straight to "chewy center"

Firewall

Servers and network devices

Tootsie Pop Network Security

Tootsie Pop Network Security: An August 2002 study performed by Asgard Group in Ft. Lauderdale, FL, found more than 150 companies using wireless networks with no security measures installed. These networks were completely exposed to both casual wireless users and hackers.

Even with all of these security systems in place, an insecure wireless network allows an attacker to bypass them, opening a path straight to the center of the Tootsie Pop, where an attacker can get past the hard outer shell undetected.

MISCONCEPTION #3:
"We'll deal with an attack it if it happens. Until then we don't have the budget."

The costs of securing a wireless network are minimal compared to the potential losses caused by a single breach. According to recent surveys by the Computer Security Institute, the number of attacks and the risks of intrusions are increasing daily.

Aside from the obvious issues, such as customer dissatisfaction, time spent restoring files, and loss of business while systems are down, management must consider the financial penalties and liabilities.

A security breach that leads to unauthorized access of data or theft of information at a public company could lead to negligence lawsuits against officers and directors for "failure to exercise due care" in protecting the privacy of its customers.

For healthcare, banking and financial services businesses in the U.S., regulatory agencies have already defined strict requirements to improve information security practices. Failure to comply may result in fines, suspensions of licenses and other disciplinary actions.

A growing trend among Fortune 1000 companies is to require vendors and partners to meet stringent network security guidelines. Failure to adhere to these guidelines may result in contract cancellations.

If your company's data, network performance and system availability are important, you should do all that you can to secure that network before damage is done.

BUILT-IN "SECURITY FEATURES" OF 802.11

Because wireless networks use radio waves to carry information, they are particularly vulnerable to unauthorized eavesdropping and access. The 802.11 wireless network protocol defines various mechanisms that can be used to help secure a wireless network. Unfortunately, in practice, many of these built-in "security features" offer little or no protection.

Service Set ID (SSID)

The SSID is a string of characters used to identify a wireless network. Although not intended for use as a security feature, some installations use it as such. For a wireless Network Interface Card (the "wireless" card in a laptop or PDA, a.k.a. the "NIC") to associate to an access point, either the SSID must be manually entered into the card, or the access point must be broadcasting the SSID.

Security Rating: Weak. Some network administrators disable SSID broadcast for security so that only cards with the correct SSID can connect to the access point. However, with certain access points, broadcast SSIDs cannot be disabled, so this option is not available.

Furthermore, anyone using an 802.11-capable raw frames sniffer can view the packets involved in a NIC associating with an access point. A raw frames sniffer program collects and displays all 802.11 packets transmitted in an area. Sniffer programs, such as *Ethereal*, are freely available on the Internet, and their use is often difficult to detect. These packets contain the SSID, which the hacker can then manually enter into his wireless card to obtain legitimate access privileges.

Media Access Control (MAC) Address Filtering

Every network card has a unique address called a MAC address assigned to it by the manufacturer. This number can be used to build filters on some brands of access points, allowing access only to the addresses on the list.

Security Rating: Weak. An intruder using a wireless sniffer can also view the frames transmitted by an authorized user to obtain a legitimate MAC address and assign it to his own wireless card. This fools the access point into thinking the attacker's card is an authorized one.

Wired Equivalent Privacy (WEP) Encryption

When the 802.11 protocols were drafted, a new encryption algorithm was developed to provide some resistance to eavesdropping. The aim was to give wireless traffic a level of privacy equivalent to a wired network. In reality, WEP turned out to be far less effective.

Security Strength: Weak to Moderate. WEP is so profoundly flawed that some security experts have suggested that it is not worth using. There are tools freely available on the Internet, such as *AirSnort*, that allow an attacker to monitor WEP-encrypted traffic on a wireless network for a period of time, and then recover the WEP key. After entering this key into his wireless card, the attacker can decrypt any traffic encrypted with this key. This attack can be effective with as little as one hour of monitoring (depending on the network traffic volume).

Since cracking WEP requires an increase in the amount of time and effort an attacker must spend, it should be enabled if no stronger encryption is available. It provides some measure of protection from less determined hackers seeking targets of opportunity and from casual unauthorized access.

802.1x/EAP/LEAP

In an attempt to shore up the weaknesses in the security mechanisms previously discussed, the industry has turned to some other protocols originally intended for wired networks referred to as the Extensible Authentication Protocol (EAP) and 802.1x, which is built on EAP.

The purpose of these protocols is to securely authenticate wireless users to the network. Before a wireless user is able to connect to the network, these protocols ensure *"you are who you say you are."*

Neither of these protocols are officially part of the 802.11 standard and are not yet supported on many vendor's products.

Lightweight Extensible Authentication Protocol (LEAP) is a Cisco-proprietary form of EAP that, in addition to improving authentication, also attempts to fix some of the problems with WEP by rotating keys to prevent easy decryption.

Security Strength: Moderate to High. There have been published reports of exploitable flaws discovered in 802.1x. More field testing of 802.1x and EAP is needed to ascertain its effectiveness for adding security to wireless networks.

If Cisco's claims about LEAP are accurate, it could provide a high level of security to wireless networks based on Cisco equipment.

HOW TO IDENTIFY AND REPAIR COMMON VULNERABILITIES IN WIRELESS NETWORKS

Many of the features and capabilities that make wireless networking so appealing can be turned into major liabilities by an attacker. However, with proper forethought and planning, a wireless network can be made fairly secure. This section presents the major security issues involved with wireless networking and countermeasures to combat them.

Weak Default Configurations (Severity: High)

Manufacturers want their products to be as easy to use as possible. If your Aunt Mary hears that wireless networking is as easy as plugging in an access point and slapping a wireless network card into her laptop, she is much more likely to buy the equipment. To this end, almost every manufacturer turns off all of the available 802.11 security features by default.

The default configurations make it trivial for an attacker to eavesdrop, gain access to the network and attack internal systems (if the access point is attached to the internal network).

Countermeasures. First, read the wireless product documentation. Under increasing pressure from industry groups and the U.S. government, manufacturers are including instructions for activating the available security options on their equipment and documenting the security risks of using wireless networking. Second, the default names, SSIDs and passwords for all wireless access devices are easily found on the Internet. This information can be used by an intruder to access your equipment, lock you out or even disable other security features. Your network administrator should change all default access-point passwords, SSIDs, names and so on to non-descriptive labels that do not identify your company or location.

Broadcast Nature of Radio (Severity: High)

Wireless networks use radio waves to carry information. Since radio waves pass through walls and other solid objects, there may be leakage of these waves outside of your facility. Due to the fact that eavesdropping on wireless network traffic requires only an inexpensive wireless card, a laptop and sniffer software (freely available from the Internet), almost any moderately skilled attacker can detect access point transmissions, intercept packets and read the contents of unencrypted packets. The passive intercepting of packets is often undetectable, so you may never realize it is even happening.

Countermeasures. The only sure way to protect from eavesdropping is to use strong encryption. As discussed earlier, WEP is not strong encryption, but it will protect you from casual attackers. If there is any sensitive data being sent over the wireless network, then stronger medicine is needed.

Encryption algorithms that have withstood the rigors of years of peer review, research and attacks are the most secure. IPSEC-based Virtual Private Networks (VPNs) using the Triple Data Encryption Standard (3DES), or the newer Advanced Encryption Standard (AES), fit this criterion.

It has been said that encryption is like a fine wine—it gets better with age. These algorithms have withstood the test of time and are considered far more secure than WEP. However, to implement an IPSEC VPN over wireless requires additional equipment and soft-

Table 6.1	How to Identify and Repair Common Vulnerabilities in Wireless Networks
THREAT	**Weak Default Configurations**
DESCRIPTION	Products are shipped with all 802.11 security features disabled.
SEVERITY	**High:** Default configurations make it trivial for an attacker to eavesdrop, gain access to the network and attack internal systems.
SOLUTION	1. Enable security settings according to the manufacturer's instructions. 2. Change default passwords, SSIDs, names and labels. 3. Do not use descriptive labels that identify your company or location.
COST	Negligible. Other than the time required to activate the available features, there is no added cost to take this primary security step.
THREAT	**Broadcast Nature of Radio**
DESCRIPTION	Radio waves pass through floors, windows and walls, potentially leaking data.
SEVERITY	**High:** Eavesdropping on wireless network traffic requires only an inexpensive wireless card, a laptop and freely available sniffer software. This passive capture of packets is often undetectable.
SOLUTION	1. Place access point antennae toward the center of a facility and use directional antennae to minimize radio leakage. 2. Use encryption so that "sniffed" packets cannot be deciphered. a. WEP b. VPNs using IPSEC c. LEAP
COST	1. Some access points offer directional antennae options (often <$100). A survey to determine leakage should also be performed by a security professional experienced with wireless sniffer techniques. 2a. WEP is a standard feature on all wireless access points and cards. It should be activated if no other stronger encryption is used.

(continued)

2b. VPNs (virtual private networks) using stronger IPSEC encryption vary in cost from $500 – $10,000+.
2c. LEAP is proprietary to Cisco. Costs will vary.
Note: An experienced security professional should be consulted to determine the most effective encryption solution.

THREAT	**Controlling Access to the Network**
DESCRIPTION	Radio wave leakage may enable intruders to access the network from outside of the facility.
SEVERITY	**High:** Wireless access points connected directly to the internal network may provide an attacker with access to unprotected servers and systems. Many networks are secured only on the outside edge with routers and firewalls. If internal systems are not fully hardened, the attacker has access to the "soft chewy" center of the network.
SOLUTION	1. Wireless networks should always be treated as untrusted networks and placed outside of the firewall, like any Internet connection. 2. When possible, the firewall should allow only IPSEC VPN connections and traffic from the wireless access points. 3. Internal servers and systems should have periodic vulnerability scans and hardening (updates, patches and other security measures).
COST	1. Network design, firewall reconfiguration and vulnerability assessments are standard security consulting services. 2. Costs will vary depending on your network infrastructure and internal security capabilities. *Consultation with a security professional is advised.*

THREAT	**Unauthorized Access Points**
DESCRIPTION	Employees may install access points without IT's knowledge or approval. Intruders may install rogue access points in a hidden location on your network.
SEVERITY	**High:** Employees generally install access points insecurely, opening up all exposures previously listed.

Rogue access points set up by intruders will be used for theft or disruption of service from outside of the facility.

SOLUTION	Educate employees about the security risks of wireless networks and create an Acceptable Use Policy that prohibits the installation of unauthorized wireless equipment. Periodic audits of the network should be performed to detect rogue access points. Audits should include a visual search and also "sniffer" scanning.
COST	Comprehensive security audits for a mid-size company range from $2,000– $30,000 depending on the size and complexity of the network. These audits can include testing network for security from the Internet, from inside the network and from alternative access including wireless access points and dial-up modems.

THREAT	**Lack of Accountability**
DESCRIPTION	A company with an insecure wireless access point may become an unwitting provider of high speed Internet access for hackers, casual users and even other companies.
SEVERITY	**Moderate:** A hacker using your wireless network to access the Internet may be difficult to identify. His activity, whether benign or malicious, will be attributed to your company. Consequences range from poor Internet performance due to heavy line utilization to lawsuits from other companies attacked via your Internet connection. .
SOLUTION	1. Limit Internet access via the wireless network through firewall rule sets. 2. Use intrusion detection to monitor all Internet and wireless network traffic for suspicious behavior. 3. Maintain logs from IDS, access points and firewalls to provide a "trail of evidence" in the event of a breach.
COST	1. Costs vary by network. 2 and 3. Intrusion detection software cost ranges from $0 for open source IDS like Snort to more than $10,000 for enterprise commercial systems. IDS requires a staff knowledgeable in the system to configure, monitor and react to alerts. Outsourced managed IDS ranges from $500 – $50,000 per month.

ware. A VPN can be described as a network of private, encrypted tunnels that use a publicly accessible network to transport data between two sites. Traditionally, VPNs have been used to allow companies and individuals to access encrypted, private data securely via the Internet. Similarly, on wireless networks, VPNs can be used to protect data transferred between a network and a wireless user.

Access point antenna placement is important in minimizing eavesdropping potential. Locating antennae toward the center of a facility and using directional antennae can minimize radio leakage, making it harder for intruders to receive packets or even locate the network.

Controlling Access to the Network (Severity: High)

Controlling access to a wired network is really a physical security problem. If an intruder wants to tap into the wired network they usually have to gain access to the facility and plug into an unused network jack or tap into a network cable. With proper controls on who can gain this level of access, this threat is minimized. In contrast, since wireless network radio waves will leak out of a facility, an intruder may effectively gain "physical" access to the network without ever entering the facility.

If wireless access points are connected to the internal network directly, then an attacker may be able to access internal systems. As mentioned earlier, many networks are only hardened on the outside edge (at routers and firewalls connecting to the Internet), and rarely are internal servers and systems hardened as fully as they should be. Thus, the wireless network gives an attacker access to the "soft chewy" center of the network by bypassing the firewall altogether. These Tootsie Pop networks are now at the mercy of almost any attack the intruder wishes to use. This can lead to a devastating intrusion.

Countermeasures. This is an authentication and authorization problem. In other words, the challenge is making sure that users are authorized, are who they claim to be and are allowed to do only what they need to do to perform their jobs. Strong encryption and 802.1x/EAP can help solve this problem.

Wireless networks should also always be treated as untrusted networks and placed outside of the firewall, like any Internet connection. The firewall can then be configured to allow only certain connections, or even better, allow only VPN connections.

Periodic vulnerability scans and hardening (updates, patches, and other security measures) should also be performed to identify and protect against vulnerabilities discovered on servers, network devices and PCs.

Unauthorized Access Points (Severity: High)

Due to the ease of installation and low cost, employees are buying and installing their own unauthorized access points on company networks. Sometimes, installation is as simple as plugging the access point into an unused network jack in the office. If an employee installs an unauthorized access point and doesn't properly secure it, he may expose the entire internal network to intruders.

Rogue access points can also be installed on your network by an intruder to be used for theft, eavesdropping or disruption of services. If this happens, you may become the unknowing owner of an open wireless network even if you never considered using wireless!

Countermeasures. Employees should be educated about the security risks of wireless networking. As part of this training, a formal Acceptable Use Policy can be created to prohibit the installation of unauthorized wireless equipment. Also, periodic audits of the network should be performed to detect rogue access points. Visual searches for rogue access points should be performed wherever network jacks are located (under desks, in network closets and computer rooms). 802.11 survey sweeps should also be used to detect any active access points.

Lack of Accountability (Severity: Moderate)

A popular trend today is wireless "hotspots," where a business with a reasonably fast Internet connection and a wireless network will allow other wireless users to "piggyback" on their connection for Internet access. A company with an insecure wireless access point may become an unwitting hotspot provider for roving users, hackers and even other companies!

While the merits of this communal generosity are beyond the scope of this chapter, there are some concerns about less-than-ethical people abusing these effectively anonymous Internet connections to attack other locations on the Internet or release the next Nimbda virus. In fact, the owner of the wireless network and Internet connection that is used to attack another company could mistakenly be blamed because the traffic will be coming from their Internet addresses.

Countermeasures. Implementing the previously mentioned security mechanisms can definitely reduce this threat. Firewall rule sets can also be used to limit Internet access from the wireless network. Intrusion detection systems, which act like burglar alarms for networks, should be implemented to monitor all Internet and wireless traffic and note suspicious traffic. And finally, maintaining accurate logs from the access points, IDS and firewalls can be useful in providing a "trail of evidence" in the event of a breach.

CONCLUSION

Network security professionals must constantly balance a business's goals with the risks and vulnerabilities when evaluating new technologies. Wireless networking delivers mobility, flexibility and productivity at an economical price. In less than two years, its convenience and price have led to widespread use and exponential growth throughout the U.S. Businesses and individuals are demanding it.

The darker side of wireless networking is its vulnerability to eavesdropping and attacks, which springs from many sources, including a lack of security awareness, user misconceptions, weak security features and insufficient network security practices. However, many of these vulnerabilities can be reduced, if not eliminated, with appropriate security technologies and practices.

Christopher Day, CISSP, is a founding partner and chief technology officer of the Asgard Group, a network security company based in Ft. Lauderdale. His specialty areas include network intrusion detection and wireless security. Day's technical achievements include creating secure extranets for e-commerce and electronic banking systems, and deploying large-scale networks to support the high-level encryption and security requirements for international banking and financial organizations. He has consulted for companies located throughout the North America, Latin America and Europe and speaks frequently at industry conferences and events.

SECURITY STRATEGIES FOR WIRELESS NETWORKS

PLANNING: Design your wireless network with security in mind. Carefully consider antenna placement and have a radio leakage survey performed during installation.

ISOLATION: Treat wireless networks as untrusted. Isolate access points using a firewall, and filter traffic from wireless networks to allow only what is needed for legitimate users to complete their tasks.

ENCRYPTION: Use strong authentication and encryption, such as 802.1x/EAP or IPSEC-based VPNs, so that intercepted data cannot be deciphered. A well-designed IPSEC VPN is arguably the best way to secure a wireless network today.

DEFENSE-IN-DEPTH. Harden the network to its core. Patch servers and desktops. Supplement your firewall and anti-virus software with intrusion detection systems. Develop security policies and procedures. Educate your users about security threats.

POLICY: Educate employees about the risks of wireless networking and issue a written policy prohibiting unauthorized installations.

TESTING: Perform periodic vulnerability assessments of your network. If possible, hire third-party security professionals to perform these assessments. Their expertise will provide a clear picture of your strengths and weaknesses.

TECHNICAL: If you are unable to perform some or all of the above, at the very least do the following to every access point on your network (consult your equipment documentation for instructions):

- Turn on WEP (use the longest key your equipment will support)
- Turn off SSID broadcast, if possible
- Enable MAC-based address filters
- Make all names and labels unidentifiable to a casual user
- Activate and periodically review access point logs
- Turn off the equipment when not in use

Source: Asgard Group, 2002

CASE STUDY:
Distribution center streamlines with WLAN

Wholesale grocery distributor Associated Food Stores (AFS) sees about 600 truckloads of groceries pass through its 600-acre distribution center in Farr West, Utah, each week. Grocery products are packed, loaded and sent on their way to more than 600 different grocers in eight states.

For many years, running the distribution center was "a matter of operating in crisis management," says Tim Van de Merwa, AFS internal logistics manager. There was little room to even evaluate efficiency, since the operation required minute-to-minute input.

A truck driver would arrive at the AFS distribution center, which used to be located near company headquarters in Salt Lake City, and go through a rather intricate check-in process. The driver would have to get out of the truck, walk up to a computer terminal near the entrance to the distribution yard and enter identifying information. The average input time was 5 to 10 minutes, though it could be longer depending on the driver's skills. Van de Merwa describes the old computer program as cumbersome and unforgiving; if one character was off, the entire entry would be rejected.

"We were expecting truck drivers to switch from their primary role to one of data entry, so that learning curve was huge," says Van de Merwa.

Truckers then had to go through another process in which they checked in with security and found out where to drop their loads. Often, the location they were told to proceed to would be full, so then they'd have to check in with security again. Some drivers would arrive and, after a long, hard day of driving, they'd say "forget the process, I'm just gonna drop it here." Or maybe they'd go through the process accurately, but with a foot of snow on the ground, they'd end up dropping it in the wrong place anyway. The result for AFS was a 40 to 50 percent accuracy range. "If we had 70 percent accuracy within that system, we felt very good," says Van de Merwa.

Since AFS operates as a co-op, there's little standardization in terms of the trucks and the cargo loads—one truck might be refrigerated, another might have no side doors, another might have roller doors. When AFS "builds a load" (puts together a pallet of grocery products) there are numerous possible configurations.

"We flow into seven states, and each state has its own regulations for height and weight, so that all comes into play as well," says Van de Merwa. "First we have to match the federal requirements, and then we have to match the store requirements—for example, some don't have docks."

The older yard-management system AFS used didn't streamline this already complicated process. Van de Merwa describes a common problem: "We'd find the kind of trailer that we needed for a specific load, and it would require us to punch into the computer, send a hossler to the location, and then it wouldn't be there. So we'd spend time going up and down rows and finally come back and say 'it's not here.' Might be there, but the driver's already spent 15 to 20 minutes looking for it in the middle of the night. So then we'd have to go back to the drawing board and look for another trailer."

Knowing where a trailer or another piece of equipment was depended on a truck driver or a yard employee logging into a computer terminal and logging the arrival or movement of a particular object. "When we started looking at the ROI, the question that was raised to me was 'How is this different than a pad of paper?' And the answer was that it really wasn't," says Van de Merwa. "We were using fancy technology, and we paid some big dollars for it, but the gain in return wasn't a whole lot different than a piece of paper and a radio."

When a reseller approached Van de Merwa with the idea of using real-time wireless location technology to manage the distribution center, he was intrigued. He and his staff spent nine months educating themselves about the wireless approach and what it would entail. What they needed was a yard-wide wireless network enhanced with location technology that would allow them to track all the vehicles and cargo moving throughout their property. AFS had evaluated a number of vendors, and finally chose WhereNet based on the company's ability to quickly provide them with proof of concept.

AFS then went through a three-phase implementation process. First, they set up a test model in their previous Salt Lake City location for six months. In the meantime, AFS was busy remodeling a new facility in Farr West. So they first outfitted part of the new Farr West center, and then added the wireless system to the final remodel. From start to finish, the implementation took about 14 months.

Though WhereNet now offers a WLAN system with wireless antennas, AFS had to lay cable for their antenna system at the time of their implementation. Van de Merwa says this part of the process was the most difficult for the company, as it involved laying five miles of cable under 12 to 18 inches of concrete.

AFS went "live" with their new wireless system in August 2001. The WLAN-based network provides AFS with real-time location and telemetry information for each of the company's several hundred trailers and yard equipment. What used to take 127 people doing data entry now involves input from just 3 to 4 people.

Now, when a driver arrives, he goes right to the distribution yard's fuel island. By the time he's done fueling, AFS managers know he's there. They also know the status of his equipment— whether it's loaded or empty, for example. The driver then proceeds to any one area of the distribution yard and drops his trailer in a drop-off spot. He doesn't have to report in, and no one has to tell him where to put the trailer. A few minutes later, a manager gets an alert about where his trailer is. A piece of equipment can enter the yard, park and be reassigned without anyone having to do anything. Before wireless, this single process was taking 15 to 30 minutes of collection and monitoring.

Van de Merwa says that AFS realized a complete return on investment after about six months. He credits this to a number of factors:

Warehouse savings: Perishable products aren't compromised while waiting to be loaded onto trailers; warehouse laborers don't have as much idle time waiting for trailers to arrive, so there's less downtime. The wireless system also transmits information about temperatures of the frozen foods and produce compartments, the fuel levels of trailers' refrigeration units, and the status of the refrigerated trailers' doors (open or closed).

Reduction of equipment: Around July, AFS sees its yearly spike in business, which often means shipping 16 extra loads with one day's notice. In years prior, Van de Merwa says he'd get on the phone and lease 10 or 15 extra pieces of equipment to deal with the increase. This past summer, he didn't have to lease a single piece of extra equipment. "We were spending $40,000 on leasing equipment—we had to lease about 45 per month in peak periods," says Van de Merwa. " The lease savings alone would pay for the new system, let alone the other ROI factors."

Less tractor usage: Over the past 18 months, AFS has gone from 92 tractors working in its distribution center down to 62. In taking that many tractors off the road, the company saves on maintenance and fuel—which has added up to about $500,000 so far.

Fast emergency restoration: When AFS experiences a power outage, the most crucial factor is how quickly they can get back up and running. In times past, if AFS had a power outage, the company was looking at a six- to seven-day recovery process. With the new system, once power was restored, AFS is back up in minutes because they had immediate visibility of all inventory—so they didn't have to rebuild from scratch the very first load. "This has been one of the most amazing and rewarding discoveries of the process," says Van de Merwa.

THE NUTS AND BOLTS

Hardware: Ruggedized OBC TELXON units (mounted in trucks), WhereTags (active wireless RF transmitters), WhereLAN (a local area network of locating access points), and the WhereNet Visibility Software Server

Applications: Wireless real-time locating system

Number of users: 500

Implementation time: 90 days (includes interfaces to yard management, warehouse management, and transportation systems)

Cost of project: Not available

Network service provider: In-house (home grown)

VARs/vendors: WhereNet Corporation

CASE STUDY:
Restaurant Saves with WLAN

While a good bowl of chili is decidedly low-tech, the manner in which restaurant chain Skyline Chili delivers it to customers is pretty high-tech. Skyline Chili restaurants have been around for more than 50 years, with most business centered in the Cincinnati, Ohio, area. The restaurants serve up a multitude of chili-centric meals to patrons in a casual setting. This isn't a typical fast food establishment; diners get china, silverware and table-side service. But the restaurant does pride itself on its ability to deliver food quickly—Pete Perdikakis, owner of three Skyline Chili franchises in the Cincinnati area, boasts that his customers can get made-to-order meals in less than two minutes. Quite a feat, considering that each of the three franchises sees 500 to 700 customers per day.

Until recently, Perdikakis accomplished this feat the old-fashioned way. A customer would walk in, sit down and give a waitress an order. The waitress would scribble down the order and then call it out to the food-prep team working behind the steam table in the main dining room. A couple minutes later, "order up," and a customer had a fresh plate of food.

The POS system worked, but there was certainly room for errors. Correct billing was dependent on busy waitstaff entering prices and tabulating bills on pads of paper. If a customer changed an order or made an addition, a waiter had to remember to record it on the bill. Wait staff had to enter the bills by hand into a cash register and then present them to the customers. Perdikakis was also concerned about "freebies"; never knowing how many complimentary meals were doled out to friends of the staff who stopped in for a bite.

Some of Perdikakis's fellow franchisees had implemented wired systems that automated the order process once a waitperson entered a customer's order into a computer terminal. "But I never saw the logic of a waitress taking an order at a table with a pen and a pad of paper, and then walking over to a terminal and inputting the order again," says Perdikakis.

"In the restaurant industry, we don't have very many ways that we can increase productivity," says Perdikakis. "Other industries can do that by changing the manufacturing process, using different materials. We can't—the biggest change in productivity for us was the drive-thru window."

Companies kept approaching Perdikakis about hard-wired terminal systems and he kept saying "No, I want wireless." An admitted "techie," Perdikakis says he couldn't figure out why his UPS driver could have a wireless system and he couldn't. After being told it couldn't be done numerous times, Perdikakis finally found a reseller who was willing to take on the challenge.

Perdikakis worked with Randy Burnette of reseller Ideal Inventory Systems to develop a system that would allow Skyline Chili's waitstaff to enter orders into handheld computers and instantaneously transmit the information to a printer at the food prep station. Perdikakis decided to set up an 802.11b WLAN in each of his three restaurants. He equipped his waitstaff with Hitachi touchscreen devices that measure about five inches by seven inches— something between a tablet-sized device and a PDA. The devices attach to a strap that can be slung over a waiter's shoulder. Perdikakis chose the devices because they allowed him to fit the entire Skyline Chili menu on one screen.

Perdikakis implemented the first WLAN order system at a brand-new Skyline Chili restaurant in late 2000. Not the best idea in hindsight, he acknowledges. "The first few weeks were hell," he says. "I wanted to put the handhelds out in the street and let trucks run over them. I thought I'd made the biggest mistake of my life."

New employees, a new manager, a new system, new devices a new store... the learning curve was just a bit too high.

But after a few weeks, Perdikakis's new store was running smoothly on the WLAN system, and when he installed it at his second and third stores in 2001, it took just a couple days for the staff to get up to speed on the new system. The learning curve is now just about 15 to 20 minutes for waitstaff to become proficient on the wireless devices, says Perdikakis.

"Managers were at first fighting it tooth and nail," says Perdikakis. "Now, if I were to try to take this away, they'd fight me tooth and nail."

Two of the Skyline Chili franchises have four handhelds, while the other has three. These numbers accommodate the maximum number of waitstaff in each restaurant. Now, when a waitress walks up to a table, she enters the order into her mobile device and receives prompts to ask questions about common "extras" (onions, no onions, for example). The waiter can then walk through an order with the customer to verify, touch the "leave table" button, and the order shows up at steam table to start to be prepared. The system prompts a printer to print out a receipt for the steam table and for the customer; the waitress can just grab the receipt from a nearby printer station and drop it off at the customer's table.

The wireless system eliminates the "human error" issues that used to concern Perdikakis—prices are pre-programmed; bills are tabulated by a computer rather than a waiter; and any extras have to be added to a customer's tab before the kitchen staff prepares them. If a customer decides to split a check, waitresses used to have to go spend 5 minutes reworking the tabs. Now it just takes a minute to reprint the check.

Perdikakis has gained additional benefits beyond the streamlined order fulfillment process within his three restaurants. The

PixelPoint software program that his system is based on integrates time-clock functions, scheduling and inventory. The inventory system has been particularly helpful, says Perdikakis. When a franchise manager runs inventory at the end of the week, the system tells her which ingredients the store ran particularly low on. "It has recipes for every item on our menu, so when a waitress presses a button for a certain order it records all the ingredients and takes them out of inventory," he says.

"You could say 'what's an extra order of cheese?' But you multiply that times 13 waitresses a day and 365 days per year, and then you're talking about a big chunk of change," says Perdikakis.

Perdikakis can track sales information by almost any factor— waitress number, table number, time of day, item sold, average check—to create a custom sales report. Based on mobile database technology from iAnywhere, the sales data can be viewed at any computer terminal in a restaurant or the back office, or on any of the handheld devices.

It's the inventory system and the improvement in bill tabulation that Perdikakis credits for his 14 percent reduction in food costs since the implementation of his WLAN system, which cost him $25,000 per store. "Now I know for sure that all my sales are being harvested," he says. "When you add in the time we were spending figuring out time cards, orders, and inventory, that 14 percent figure is way low."

THE NUTS AND BOLTS

Hardware: Hitachi HPW 600 handheld tablets/Proxim WLAN

Applications: PixelPoint 2000

Number of users: 12

Implementation time: Seven months for three store implementations.

Cost of project: $75,000

Network service provider: Proxim 802.11B WLAN

VARs/vendors: PixelPoint, iAnywhere, Ideal Inventory Systems

CASE STUDY:
Medical Center Gets Data to Doctors Faster with WiFi

With a 467-bed hospital and 23 physician clinics, NorthEast Medical Center's 3,200 employees have their hands full. The Concord, North Carolina, medical center recently put a little more time back into the hands of its employees with a new 802.11b wireless network and several healthcare-specific applications.

In 2001 NorthEast Medical installed an 802.11b wireless data network throughout its 60-acre campus. Starting in 2002, the center began adding software applications that give physicians and nurses in the hospital and the physician clinics access to data including medical records, workflow efficiency systems and wireless prescription services.

"The number one driver behind our wireless project was patient safety," says Keith McNeice, NorthEast Medical's vice president. "This doesn't imply that we were unsafe, but it's more a matter of recognizing that by giving clinicians access to the most up-to-date information at the point of care, we improve patient care."

NorthEast Medical uses a MercuryMD application to give physicians handheld access to patient records—where a patient is in the hospital, lab results, medications, and reports from various departments such as radiology and pathology. The application will run wirelessly or via a docking station. Doctors can choose to use either a PocketPC- or Palm-based device, but only the PocketPC devices can be used to access data on the wireless network.

"We chose this two-platform approach because we had a large number of doctors who were already using some form of PDA," says McNeice. "We had always encouraged them to do so in the past, and had given Palms to our doctors in the past to demonstrate value, so we wanted to allow our physicians to continue using the device form they preferred."

However McNeice notes that his organization has drawn a line and said that it won't support the Palm device in a wireless environment. And the medical center's future development is all going to be based on the PocketPC platform. "But it worked well for us to be able to introduce the MercuryMD application without worrying about prejudices in terms of device preferences," says McNeice.

Since the MercuryMD system hasn't been in place for an extended period of time, McNeice only has anecdotal evidence of its efficacy thus far. But some doctors have told him that the application saves them an hour a day. Others say it isn't necessarily saving them time, but its getting them important patient information much faster, so they're able to make corrective changes in the course of a patient's treatment sooner. The application is saving nursing staff time too, says McNeice, since the nurses don't have to look up information and pull files for the doctors.

In physician clinics, NorthEast Medical has implemented a program that allows doctors to write prescriptions on PocketPC-based handheld devices. The program can then access the wireless network to cross-check prescriptions against a patient's previous medications, allergies and the formulary that the patient's insurance company supports. Doctors can also wirelessly transmit a prescription to a patient's pharmacy with one click on their handheld devices. The prescription arrives at the pharmacy via fax with a fully legible electronic signature.

In the medical center's old paper environment, a physician wouldn't have this info at her fingertips—a patient might get the prescription that the doctor normally prescribes, take it to his pharmacy and then find out that his insurance company doesn't approve the particular formulary that he's been prescribed.

NorthEast's third application from Bridge Medical makes use of laptop PCs on mobile carts in nursing areas. The wireless program is based on barcodes—its primary function is to improve patient safety in the administration of medications. All employee name badges are barcoded, as are patient name bands and all medications. A nurse uses the mobile laptop to pull up a patient's medical profile and find out what medications she needs. The nurse then enters the patient's room with the medication and uses a scanner to enter the patient's barcode and the medication barcode. The system verifies that the medication is correct, then enters "received" into the patient's record. In the hospital's old environment, all of this information would have to be handwritten into a patient file. The program also interfaces with NorthEast's pharmacy system to reconcile inventory

"It automatically charts in the system that the medication was given, who it was given to, and when and by whom," says McNeice.

The nursing staff at NorthEast is currently researching usage of the Bridge Medical application to try to determine the time and cost savings for the hospital. Anecdotally, some of the more experienced nurses say it's saving them in more than half an hour each shift. "We had one instance during training the doctors on the MercuryMD system where a doctor got a lab result on his handheld and was able to immediately pick up the phone, call the nurses and change the care plan for a patient," says McNeice.

McNeice says that he's had no problem garnering enthusiasm among staff for the mobile projects. "Part of that is in how you sell it, how you pick the right app at the right time," he says.

Picking the right applications was a huge part of the process for NorthEast. McNeice says that they started by going on site visits to many of the vendors who specialize in the healthcare market. They then chose a few vendors to come in and present to the hospital's clinic managers. The medical center's key criteria for selecting applications was integration.

"The core hospital system drives all our other systems," says McNeice. "So one of our big questions for vendors was, 'Tell us what hospital systems you've had experience interfacing with and tell us how that interface works.'"

NorthEast has a relatively small IT staff, but it's composed of a high percentage of IT professionals with a clinical background. The implementation process for all three of the hospital's wireless applications was led by former nurses who work in IT. "When they go and talk to their peers in nursing or to physicians, there's an instant credibility there that a technical person would not have in terms of relating to the workflow and clinical relevance," says McNeice.

NorthEast still provides classroom-type training for the applications, but for the day-to-day training and support, the medical center has appointed a few key staff who can act as educational coaches for their peers. "We already have nurse educators on staff who are responsible for ongoing training and patient education, so these are usually the people who are designated," says McNeice.

Of course, in a medical environment, security issues are more serious than in many other business settings. "We have all sorts of different layers of security on our landline and our wireless networks," says McNeice. "And security was a big issue when we evaluated the applications that were available to us."

NorthEast also bolsters its security by employing outside security firms to test its network and determine where it might have weak points. Patient data is never stored on the handheld devices—the devices only interact with the hospital system. Once data is downloaded to the device, it's encrypted. And there's a self-destruct feature so that if someone tries to break into the device, after a certain number of tries the data is deleted.

THE NUTS AND BOLTS

Hardware: PocketPC-based handheld devices, IBM laptops, Symbol barcode readers, HP Proliant servers for MercuryMD and AllScripts, MercuryMD infrared synch stations

Applications: Internet Explorer for access to clinical reference services, MD Coder for charge capture/coding, AllScripts for outpatient prescription writing and drug reference, MedPoint from Bridge Medical for barcode application

Number of users: 150

Cost of project: Not available

VARs/vendors: MercuryMD, Bridge Medical, Allscripts Healthcare Solutions, Mobile Design Technologies, Extended Systems, Symbol Technologies, HP, Toshiba

Chapter 7

Communicative Machines

Sure, wireless applications can save time by helping people move faster. Isn't it even better when it can free them altogether from routine tasks? That's the promise of machine-to-machine communications, also known as M2M.

After all, in many businesses the majority of workers are glorified nannies, doing little more than keeping an eye on the machines that perform the actual work. On the factory floor, they trudge from meter to meter; in the field, they sometimes travel for hours to check on a pump or valve out in the middle of nowhere.

Enabling the overseer to remotely monitor a machine using a desktop computer offers a world of time- and labor-saving advantages. Even better is letting the machine report directly to a central monitoring and management system. This latter strategy gives the human supervisors a higher-level understanding of systemic operations and an improved ability to forecast and plan.

On the consumer level, communicative machines with wireless connectivity let homeowners monitor their houses while they're away, control home systems such as heating and cooling to respond to changes in weather, and take advantage of off-peak prices for energy and water use.

Communicative machines can make field-force workers more productive, help manufacturers and utility companies balance loads and avoid outages, help retail store clerks serve customers better, and let homeowners feel like the Jetsons.

M2M communications include:

✳ Remote sensors that allow monitoring at a distance

✳ Remote control of machinery and systems for industry and for the home

✳ Short-range sensor-to-sensor exchanges using Radio Frequency Identification tags

✪ FIVE HOT MARKETS FOR COMMUNICATIVE MACHINES

While any company that relies heavily on machinery may find benefits in connecting them to information systems, the following five sectors are leading the way.

Utilities Take Control

When the summer sun comes up, the air conditioners go on, and power supplies get stretched to the limit. Blackouts are more than public relations disasters. Business screeches to a halt, and enterprises may lose valuable data as well as staff time. Manufacturers may not be able to fill orders or may have to dump perishable materials.

Utilities have an even more important responsibility to customers such as hospitals and health care facilities, which need as much warning as possible so that they can switch over to emergency generators. Utility companies have found that anything they can do to reduce demand during peak times can have a big payoff: They avoid paying premium rates to energy producers; they avoid the fines due when they can't meet the service level agreements they've entered into with business and manufacturing customers; and they avoid the huge fines levied by the government.

Getting users, whether commercial or consumer, to reduce non-essential usage to avoid a brownout is often as easy as asking. Utilities traditionally get the word out by public relations efforts such as radio ads reminding consumers to be energy-conscious and by telephoning large customers to warn them when power supplies are low. But sometimes the plant, operations or energy manager has just minutes to drive down demand enough to avoid a blackout.

Automated notification systems can play a huge role in energy demand management. Instead of the energy manager calling or emailing commercial users, she can use a simple interface to instantly notify all critical users via whatever communications protocols work best for them: email, telephone, fax, pager or cell phone. Most often, the preferred communication methods for such time-critical alerts will include an automated voice or text message to a cell phone.

To maximize their benefits, such systems should include the ability to send interactive announcements, so that the customer can acknowledge receipt of the alert and perhaps provide information about its curtailment plans. This capability can pay off for the utility by reducing customer claims that they didn't receive the curtailment alert.

Southern California Edison rolled out such an automated alerting system in March 2002, using software from Utility Data Resources. SCE uses the system to comply with government regulations and notify key customers and public agencies of rotating outages or other emergencies. Commercial customers that sign up for SCE's Interruptible Rate Plan agree to either curtail usage of designated systems upon request or to pay a premium to keep their machines humming. An SCE spokesperson says Interruptible Rate customers were "heroes" during the summer 2001 energy crisis, responsible for avoiding several major rolling blackouts.

Such systems also help utilities improve customer service. They can offer key customers the ability to manage their energy use. For example, a manufacturer could cut its energy costs if it could get access to real-time pricing information. Before beginning an energy-intensive manufacturing process, the operations manager could log on to the utility's Web interface to see the current and projected costs of the electricity that would be used. If it was too high, she could defer the process until the swing or graveyard shift, potentially saving thousands of dollars. The benefit that wireless adds to these systems is the ability for the alert recipients to be notified on multiple devices, including the cell phone, which tends to be always carried.

Companies such as Utility Data Resources (UDR) have begun to offer applications that give commercial and industrial customers the ability to use improved information from specialized meters that provide what's known as "interval data recording." Instead of simply metering total energy usage on a constant basis, these smart meters report usage at small intervals, typically every 15 minutes.

Commercial smart meters can be read over regular telephone lines or via the cellular system. Wireless is useful when the meter's location makes it difficult or too expensive for the telephone company to run a landline, for example, deep inside a power plant or at the edge of the substation property.

Such fine-tuned information can help businesses identify trends in usage so that they can modify their operations to take advantage of pricing variations. UDR also can send the interval information to the business's internal energy-management systems, which can be set to automatically turn down non-essential uses, such as hot water heating or lighting, when the company's total usage comes close to jumping to the next pricing tier.

As in many mobile and wireless applications, the real payoff comes not in incremental savings on the cost of energy, but in adding this data source to the company's ERP systems, where it adds greater visibility into operations and the cost of doing business.

Utilities are also experimenting with bringing this same level of communication to their consumer bases. Essential to these endeavors is the ability to change settings while not on site. Remote control via wireless device lets the homeowner voluntarily turn up the thermostat for the home air conditioning system, even if he's at work, for example. In three pilots, Carrier worked with Long Island Power Authority, Puget Sound Energy and New Power to let customers wirelessly adjust air conditioners and water heaters. SkyTel added integral two-way pagers to the thermostats, which let the utility companies set back the temperature during peak hours. Consumers could also remotely control the thermostats through the utilities' Web interface, using a desktop or wireless Web browser.

The goal was to see whether, if consumers could control appliances remotely through a wireless Internet connection, they would operate the machines at off-peak hours, or save money by adjusting the heating or cooling systems when the weather changed. The wireless connectivity was key: First, it made installation much easier; otherwise, a separate landline or connection to the home's existing line would have to be made. Using the paging network lets the utility simultaneously broadcast setback orders to entire service blocks, instead of having to dial in to each home individually. (For more information about the Long Island Power Authority project, see "Case Study: Wireless Thermostats Save Power on page 193.")

A remote-controlled thermostat would benefit consumers in other ways, too. For example, it's a waste of energy to keep the heat on while the family is away during the day. But in very cold climates, homeowners must weigh the energy savings with the danger of pipes freezing or pets and plants suffering if the day turns much colder than expected. With remote control, they can easily monitor and make adjustments if necessary.

Safety and Security

A new market is developing for products that trigger alerts when home safety system alarms go off. While burglar alarm monitoring companies have long picked up the telephone to call customers on their cell phones, automating the process and supporting multiple devices, including two-way pagers and PDAs, greatly enhances the benefits. For example, Mayday Mayday is a San Francisco company that began selling a product that connects to a home PC and helps customers differentiate an alarm caused by kitchen smoke from a real fire. Smoke and Carbon Monoxide Alarm Messenger detects the sound of a home alarm and automatically reaches the homeowner according to the communication preferences she's preset.

When Alarm Messenger activates, it connects to Mayday Mayday's servers. The servers then refer the message to Envoy World-Wide, a company that specializes in multi-modal communications. Envoy translates the message and sends it out in the requested formats, which include automated voice on landline or mobile phone, SMS, WAP, email, pager and fax.

The advantage of this automated system is that the messages are interactive. The customer can automatically connect to her home or the local fire department. By calling her house first, she can find out if her teenage daughter has burned the toast; if there's no answer when someone should be home, there may be a real emergency and she can immediately reach the fire department.

Manufacturers have been tinkering with the idea of wireless remote control of home appliances for quite some time and, again, safety seems to be the most compelling selling point.

On the national level, radio frequency identification (RFID) tags are being used to tighten security at military bases by letting security staff monitor the movements of personnel and equipment. For more about how government and public safety organizations are using RFID, see Chapter 8.

Internet-Connected Home Appliances

In October 2002, LG Electronics unveiled its long-awaited multimedia refrigerator. The stainless-steel LG Internet Refrigerator, retailing for $7,900, has an integral touch-screen that lets users check email, watch television and download music files from the Internet, through a modem or high-speed Internet connection. Its owners can also log on to a Web site when they're away from home to see the to-do lists or schedules they've created on the appliance.

Critical response to the multimedia capabilities was tepid; it's difficult to see how the higher price of the product translates into real benefits for the user. Many manufacturers have toyed with letting cooks download recipes and get cooking instructions.

Both Whirlpool and LG have even considered adding a barcode scanner to the fridge. The idea was that the user would swipe products across a barcode scanner as they took them out. The appliance would keep a running tally of what was on hand, warn the user when a product became out of date or was gone, and even, someday, automatically order replenishments. Consumers will have to change deeply ingrained habits in order to take advantage of such technologies. Yet, aside from microwaving, the act of cooking itself remains one of the most traditional home activities.

A little-noticed feature of LG's model is the one that could really pay off for the company and, potentially, its customers. The Internet Refrigerator comes wired with a component that automatically contacts LG's service division if there's a malfunction. (LG will only service the actual refrigeration unit; customers will have to deal with their ISPs if they have problems with slow downloads.) LG will call the customer within 24 hours to schedule a visit to check and repair the fridge.

There is real opportunity here for appliance manufacturers' service departments. This system has three strong benefits to manufacturers: First, the repair request automatically goes to the manufacturer or dealer rather than a competitor. Second, repairs are simpler because the machine hasn't broken yet. Therefore, the repairperson spends less time on the premises, saving the company money on each repair call. Finally, the self-diagnosing feature gives the manufacturer the ability to sell a more expensive warranty, based on the guarantee that the machine will never go out of service.

The rise of home automation may make machine-to-machine communications more widespread. Networked home systems are becoming common in higher-end new home construction, while integrators specializing in home systems have made it easier to retrofit. Home automation connects various systems in a house, including heating and air conditioning, lighting, security, irrigation, and even entertainment. They are all controlled through a central processing unit and connecting to the Internet through a residential gateway.

While most home automation systems will run on wired broadband networks, some elements will no doubt connect wirelessly, especially as prices for wireless LANs fall. (For more information on WLANs, see Chapter 6.) For example, a wireless node in a detached garage might connect a motion sensor to the main security system and connect a separate heating unit to the main residential control center.

Vending Machines

Soft drink route sales drivers spend a large chunk of their work lives tending to the needs of their vending machines. They stop at each machine to collect money, check and replenish the inventory and make sure they're working well. A good driver gets to know the machines—which ones tend to get coins stuck, which ones go through all the iced tea before the cola. Still, the driver can't take a chance and skip a scheduled stop and risk missing an anomalous event that needs attention.

Vending machines that report conditions to a central location can make drivers more productive by letting them plan their routes more efficiently, eliminating unnecessary stops. Malfunction reports can ensure that the driver is prepared to fix the problem. An alert when the machine's temperature exceeds a predetermined limit can speed the repair technician, reducing lost revenue.

Communicative vending machines can help the beverage distributor understand sales trends and uncover consumer behavior—for example, if certain designs on the machine seem to boost sales of a particular product. This data can help the beverage maker's sales force in its efforts to get more slots or better placement in the machines if it can show a history of strong sales.

Deloitte Research foresees a time when vending machine companies could gain additional revenue by using their machines to deliver third-party advertising, in the same way that many automatic teller machines now do. With a wireless channel, the machine could receive a constant stream of realtime text ads, and even download rich media ads overnight for display during the day. Deloitte has identified the high cost of connectivity as a barrier to the development of communicative machines. The cost would have to plummet to make selling and managing this kind of ad worthwhile.

Retail

When a shopper at the exclusive SoHo clothing retailer Prada is ready to try on clothes, she enters a sleek glass-fronted dressing room; with a touch of a button, she opaques the glass wall. As she hangs items in a "smart" dressing room closet, a touch screen lets her choose from a wealth of intriguing information about her chosen garment, including video clips of models wearing the outfit, color swatches showing all the hues available, and what accessories the designer suggests.

These smart closets uses RFID technology to match the garment with information in the retailer's database. RFID uses tiny transponders that communicate to a reader using the radio frequency portion of the spectrum. The reader can connect with the transponder over short distances, typically no more than six feet. Transponders can carry about as much information as a barcode; but, unlike barcodes, RFID tags don't need to be scanned to be read. They automatically transmit their information, and multiple tags can be read at once. Data on an RFID tag can include product information. Tags can interact with databases; they can also trigger an action, such as unlocking a door.

Sales clerks equipped with handheld devices with integral RFID readers can use them to instantly find out what other sizes of an item are available, or how many total units are in stock. RFID has great potential for retailers. An RFID system can do the following:

✳ Automatically tell a manufacturer or merchant when supply of a product on a store shelf is low.

✳ Let shippers quickly identify the contents of a pallet.

✳ Link a customer with a database giving extensive information about a particular product.

✳ Allow shoppers to check themselves out.

Unfortunately, the utility of RFID is counterbalanced by the cost of the tags. Most RFID tags still cost more than a dollar. The increasing availability of reliable lower-cost tags could completely change the way products are handled from the factory through the cash register.

Mass retailers Gap and Target reportedly tested RFID technology in 2002 for supply chain and inventory management. In November 2002, razor and blade manufacturer Gillette announced an order of 500 million RFID tags at a reported cost of less than ten cents each. Gillette planned to embed the tags in high-end razors to better track merchandise from the warehouse to its point of sale. If this trial succeeds, RFID tags could soon replace barcodes as the major product identifier.

✪ WHAT'S NEXT?

While the technology exists to enable all sorts of valuable machine-to-machine communications, this part of the mobile industry has a big hump to get over: connectivity costs. Mobile network operators, especially in the U.S., are stuck on consumer pricing. Whether they're charging by the minute, the packet or the SMS message, the costs are too high to make sense for many machine implementations, according to Paul Lee, a director with Deloitte Research. Lee points out that SMS is the best way at this point in the industry's evolution to send information to and from machines, but at an average cost of ten cents per SMS, "It could get pricey," especially for such leading-edge initiatives as sending advertisements to a vending machine.

Lee says network operators should offer special pricing schemes for non-human mobile communications, including volume-based rates and tiered pricing for variable service levels. In return for cheaper rates, machines could communicate during the late night and early morning hours, a time when network capacity is typically underused.

That's not at all the only hurdle for the M2M industry. Standards for data interchange is another stalling point. The vending machine industry has the Vending Industry Data Transfer Standard (VIDTS), maintained by the European Vending Association and the National Automatic Merchandising Association in the U.S. The latter organization is working to extend standardization to reports, an industry-specific system of accounting criteria and an industry knowledge base of best practices.

The home automation industry is working on its own set of standards, revolving around communication with the home gateway, which will, in turn, take care of managing IP-based communications over the Internet.

These all are important steps toward the creation of a broader marketplace. This segment will tend to be highly vertical; the needs of a vending machine company, a military unit and a retail store are much more divergent than those of salespeople selling different types of products. Therefore, hardware and software vendors will need to create highly targeted applications. However, the closer they can get to interoperability and standard offerings, the more potential customers will feel that they have enough reliable choices and comparisons to take the leap.

Finally, to fully exploit the potential of M2M, companies must change their business processes to a much greater extent than for other types of wireless applications. A field-force worker is already making reports and entering them into a database; mobile applications let her do it more efficiently. But a utility company or clothing retailer that enables its physical infrastructure to report what's going on can be faced with a flood of information that it's unable to make sense of or use.

For example, Dr Pepper/7 Up found its test of Isochron's Vend-Cast application "fantastic, by all measures a success," according to company spokesperson Michael A. Martin. The technology worked really well and the company found the information on brand performance to be potentially invaluable.

But Dr Pepper/7 Up's vendors, the companies that actually own and operate the vending machines that sell Dr Pepper and its other drinks, said the cost of VendCast was too high and that there was more data than they could possibly use.

Just as handing a mobile worker a handheld device doesn't ensure greater productivity, communicative machines may not pay back their overseers unless they have something useful to say.

There are thousands of possibilities for machine-to-machine communications. Some of the M2M the applications that Deloitte Research has identified as most promising in the near future are shown in Table 7.1

Table 7.1 M2M applications for selected vertical sectors

Sector	Applications
Advertising	▪ Remotely updating digital advertising displays, located in various places from parking meters to billboards. For example, a billboard for a supermarket could advertise its latest offers, updated hourly, on a scrolling display.
Agriculture	▪ Communicating soil, temperature and humidity data to a central monitoring unit to provide early notice of adverse conditions. Early notice of conditions, such as frost for sensitive crops (e.g. grapes or tobacco), could allow a harvest to be saved.
Automotive	▪ Alerting emergency services automatically in the event of an accident (e.g. when the vehicle's airbag is deployed).
	▪ Providing drivers with in-vehicle information on traffic, maps, news, local restaurants, etc.
Healthcare and Pharmaceutical	▪ Remotely monitoring patients—especially infants, the chronically ill and the elderly—allowing them to spend less time in the hospital. This would improve patient comfort and reduce costs.
	▪ Remotely gathering drug trial data, leading to faster and more frequent feedback than with hospital-based data gathering.
Financial Services	▪ Authorizing credit card transactions in locations without fixed connections, e.g. trains and temporary exhibition centers. This could reduce the quantity of fraudulent transactions.
Media	▪ Wirelessly transmitting press photographs (low to medium resolution) where immediacy is more important than image quality. This could be based around digital cameras with embedded mobile capability.
Retail	▪ Maintaining and programming vending machines. Machines could alert stock clerks to low stock levels and technicians to mechanical breakdowns. Prices could be updated remotely.
Transport	▪ Tracking the location of cargo and hired vehicles.
	▪ Remotely controlling vehicle functions (e.g., disabling the engine of hijacked cargo vehicles or rental cars).
Utility	▪ Automated meter reading in areas of light population. Deployment would work best in multi-service offerings, where the utility provides a range of services (e.g. security) in addition to gas, water and electricity.
	▪ Transmitting weather information wirelessly from remote monitors. This could be cheaper than providing a landline connection.

Source: Deloitte Research, 2002

Reinventing Machine Communications
By Paul Lee

The ultimate impact of the embedded mobile application will depend on the ambition and creativity in its design. Deloitte Research has identified three distinct approaches to mobilizing the machine.

In its simplest form, embedding mobile communications in the machine links it to the outside world, perhaps for the first time, or perhaps in a more cost-effective manner. In itself, this approach—what we refer to as *mobile enablement*—can be powerful; but it's only the beginning. A more advanced approach—*mobile re-invention*—uses a machine's mobile communication capabilities to transform business processes, literally changing the way work is done. The most powerful approach—*mobile discontinuity*—uses embedded mobile in a way that drastically alters the business landscape, creating opportunities for business model innovation and new sources of revenue. The potential return on investment and quantity of value generated increase with each level, but so do the risk and complexity.

MOBILE ENABLEMENT
The majority of applications will fall into the mobile enablement category, particularly in the early stages of the market. Such applications require little investment, but their potential returns are unlimited.

Mobile enablement provides a new communications link where none previously existed, or offers a lower-cost alternative to conventional solutions such as a fixed phone line.

Good candidates for mobile enablement include machines in the following scenarios.

- **Used while they are in motion**, such as vehicles and portable electronic equipment. For example, embedding mobile into a credit card authorization terminal would allow online transaction authorization in moving locations such as trains and buses. This would give travelers more payment options without increasing the risk of fraud.

- **Frequently moved between locations**, such as projectors, computer equipment and industrial diggers. Mobile enabling such devices—especially those that are extremely valuable and easily lost or stolen—would allow location tracking and improve their security.

- **Frequently moved within a location**, such as cash registers in a retail store. Embedding mobile would make it easier to reconfigure the floor layout and move cash registers around, without needing to worry about the presence of a phone jack.

- **Located remotely but strategically**, such as roadside information panels on a major highway. These devices can be difficult to access with fixed lines, but are generally within the coverage area of a cellular network.

- **Located in areas where there is difficulty in obtaining fixed connections**. Even in developed countries, it can take several weeks to complete the installation of a fixed connection, but only minutes to install a mobile connection.

MOBILE RE-INVENTION

With *mobile re-invention*, embedded mobile technology is used to dramatically improve information flow and transform business processes. Processes that previously involved periodic checks or reports can be virtually eliminated. Processes that relied on prediction and guesswork can be replaced by solutions that provide immediate responses to real-life events. Tasks that required face-to-face visits can be addressed remotely.

Examples of mobile re-invention include the following.

- Equipment that sends an alert when it needs maintenance, rather than requiring a technician to make periodic check-ups. This would save on unnecessary visits, and reduce the delay for completion of repairs.

- Mobile-enabled meter readers that allow utility companies to monitor and set controls on individual home and business consumption, rather than relying on blanket brownouts and unenforceable watering guidelines.

- Mobile-enabled automobiles that could be alerted and pre-qualified in the event of a product recall on defective parts. This would help car manufacturers resolve the problem faster— enhancing consumer safety and reducing liability risk—while simultaneously avoiding the major expense associated with a full recall.

- Programmable electronic devices that can be maintained and repaired remotely.

MOBILE DISCONTINUITY

The most powerful application of embedded mobile is to enable new business models, for example, by using positioning information from vehicles to feed traffic information services. The quality of a traffic information service depends on the quantity and quality of data sources. The most accurate traffic information services today are typically based on data obtained from traffic beacons located at the side of major roads. A major investment is required to set up such a system.

Yet the positioning systems that are increasingly being deployed in cars, combined with embedded mobile technology, could provide a far more powerful information-gathering service. Vehicles could transmit their speed and position periodically via a built-in mobile modem. If the car's speed dropped, indicating a traffic jam, data would be sent more frequently. The more congested the area, the greater the flow of information. Thus, during summer months, traffic information for coastal areas would automatically improve due to the higher concentration of vehicles. A service based on this approach would only be possible with embedded mobile.

As the embedded mobile market evolves, potential applications will proliferate, and the opportunity for profit will be substantial. Those organizations that early on are prepared to embrace its unique challenges and opportunities will be best positioned to reap its eventual rewards.

Paul Lee is director of mobile and wireless at Deloitte Research. In this role, he is responsible for developing Deloitte's point of view on key trends in the mobile and wireless sectors. He's a frequent commentator on business television and is a guest lecturer at London Business School.

CASE STUDY:
Wireless Thermostats Save Power

The Long Island Power Authority (LIPA) is a municipal utility company serving Long Island, New York. In April 2001, it launched LIPA Edge, a program created by heating and cooling equipment manufacturer Carrier to help utilities manage demand (and sell more Carrier products). After installing air conditioner thermostats with a wireless Internet connection, the utility has the ability to take control of participating customers' air conditioner thermostats for short intervals during times of peak demand.

The program, for residential customers with central air conditioning, is voluntary. Participants receive a programmable thermostat with wireless modem, with a retail value of $200, plus free installation. To sweeten the deal, customers also get $25 cash when they sign up. The ComfortChoice system provided by Carrier consists of a programmable thermostat and a wireless modem that's mounted near the furnace or attic air intake. Carrier also provides Web-based software with two separate interfaces, one for the power company and one for its customers.

LIPA uses its ability to control thermostats to optimize its electric system. By shaving off a tiny bit from each household's usage, it can save electricity and thereby avoid outages. The LIPA system can either temporarily adjust the air conditioner's compressor or actually turn up the temperature setting a few degrees. When the extra power is no longer needed, the system returns the settings to the customers' preferences. The utility promises customers that it will adjust the air conditioning system no more than seven times per summer and only between the hours of 2 PM and 6 PM.

When LIPA does take control of the air conditioners, the digital screen on a customer's thermostat states that it's in "curtailment." The screen also tells the owner know how much time is left before it goes back to the preset temperature. According to LIPA, the temperature in the house typically increases no more than three degrees during curtailments. Customers are free to manually override the curtailment.

In the summer of 2001, LIPA activated the system just four times, and saved two megawatts total of electricity during each curtailment. By the end of 2002, LIPA had installed over 18,000 wireless thermostats and enrolled 19,500 customers in the program. In the summer of 2002, there were about four curtailments. According to Dan Zaweski, director of energy efficiency for LIPA's distributed energy programs, LIPA saved around one kilowatt per hour per household during each curtailment. That translates to roughly 18 to 19 megawatts of energy saved per curtailment. "We're actually saving more than what we originally expected to per household," Zaweski says.

Zaweski says that reassuring customers that they could override the curtailments was important to getting customers to enroll. LIPA also provides a Web-based interface that lets its customers access their wireless thermostats remotely. While neither Carrier nor LIPA sees end-user remote control as a key feature, it's certainly an added perk for homeowners. "This is the first time any of our customers have had the ability to interact with their homes through the Internet," he says. At this point, less than 25 percent of customers take advantage of their own ability to adjust the thermostat while they're away.

LIPA considers the program so successful that it also added a program to provide wireless remote control for customers' swimming pool pumps, which boosted its total power savings through curtailments to 22 megawatts each.

THE NUTS AND BOLTS

Hardware: Carrier ComfortChoice thermostat, Motorola Reflex pager

Applications: Enterprise Energy Information Management System

Number of users: 18,000

Implementation time: Ongoing

Cost of project: $14 million when fully implemented

Network service provider: SkyTel Telemetry

VARs/application vendors: Silicon Energy

CASE STUDY:
Rail Line Tracks Gate Operation

Montana Rail Link (MRL) is a regional railroad with more than 900 miles of track running through vast tracts of wilderness. Its 2,100 freight cars and 120 locomotives serve 100 stations in Montana, Idaho and Washington.

MRL's highway grade crossings automatically open when a train approaches, then close again when the train has passed. However, the company experiences as many as five crossing activation failures per day. Luckily, most of these involve the gates being stuck in the down position. Such failures may be caused by improper battery voltage or a commercial power failure. While a crossing grade that's stuck in the closed position isn't a safety hazard, it's inconvenient to motorists and could impede emergency vehicles speeding to a call.

Much more dangerous, though infrequent, is a grade crossing gate that fails to lower when a train approaches. When this happens, the approaching train must stop, send a person to flag traffic to a halt, cross the highway and then stop the train again so to pick up the person with the flag. This process wastes time and delays the train.

Often, a crossing gate might be malfunctioning for some time before MRL finds out about it. "There's no other way to know that a crossing is malfunctioning unless a train crew saw it, or a passerby or the police," says Ray Krenik, MRL's manager of control systems.

In 1999, MRL began to add wireless remote monitoring systems to its new installations. The wireless connectivity was essential because the monthly service charge for a traditional telephone line at each crossing gate was way too high. Also, most calls to the main dispatch center in Missoula would have been charged as long distance.

Instead, MRL's ScadaNET Network system uses standard Internet-accessible remote terminal units (RTUs). The communications system, which works over the cellular control channel instead of the voice channel, integrates the cellular telephone network, the Internet, email, faxes and pocket pagers.

The system has a Web-based interface where MRL adds information about each crossing, such as its location, and sets up the monitoring rules. The system can report whether the power at a gate is on or off, check voltage of the backup batteries, and report multiple types of malfunction, such as a gate that's only partially up. While it's capable of sending alerts to a variety of mobile devices, most go via email to the desktop computer at the dispatching center, because this rural area doesn't have full pager network coverage. The dispatcher automatically broadcasts alerts and all clears out to all appropriate personnel.

Krenik says that MRL is installing the wireless remote monitoring system gradually because cash is tight, but "it's definitely worth it." In fact, the railroad has begun to use the cellular RTUs to monitor other systems. For example, in one of its tunnels RTUs monitor the power supply to an exhaust fan, while another keeps tabs on heating units that melt ice and snow.

THE NUTS AND BOLTS

Hardware: LaBarge CTU-08

Applications: ScadaNET Network

Number of locations monitored: About 160

Implementation time: Not disclosed

Cost of project: $1,500 per unit plus $5/month connectivity charge

Network service provider: Numerex Cellemetry Data Service

VARs/application vendors: LaBarge

Chapter 8

For the Greater Good

Wireless is ready to roll for government agencies and public-safety organizations, due to a combination of heightened need and technological readiness. The Department of Defense is one huge infrastructure for field-force deployment, and it has arguably the broadest and least centralized set of assets of any public or private organization in the world. At the same time, technology to manage logistics and personnel is sorely lacking. From the Army to the Navy to the Justice Department to the FBI, assets and information are managed with a welter of mostly antique hardware and software systems.

With this century's increase in terrorism at home and overseas, government, security and peace-keeping agencies have an urgent need to increase their "data velocity," the time it takes for information to move from the field worker who gathers it to central databases where it can be plotted against that from other sources. With a wireless mobile device, a police officer evaluating a suspicious vehicle can not only query the stolen-vehicle database but can also report the vehicle's description and circumstances to other agencies such as the FBI, which may be searching for just such a vehicle.

The heightened need for security calls for a higher set of standards for tracking assets and a similarly enhanced ability to report and disseminate information. Yet many government field workers, from police officers to nuclear plant inspectors, are as paper-dependent as those in the private sector.

The public sector tends to lag behind the private in technology adoption, and most organizations have just finished installing basic business automation and enterprise resource planning systems. With these systems in place, these organizations can begin to move forward and reap the benefits of wireless applications. Their trailing edge position will help technology planners maximize their investments by learning from first movers' mistakes.

In some ways, military, police and public-safety personnel are similar to other field workers: They're highly mobile, and their core duties don't include information processing. They often work in rugged conditions, and they have a strong need for information housed at headquarters. In addition to the improvements in information and asset management mobile data applications provide, government field-force workers will experience the same increases in efficiency and productivity as those in the private sector.

While they can greatly benefit from many standard field-force applications, there are some ways in which public sector field forces are unique. For one thing, the stakes are higher.

✳ Getting information fast is crucial. While a repair mechanic may feel frustrated if his wireless connection is slow and he can't get the schematic he needs to repair a machine, it may be a matter of life or death for an emergency medical technician to get a map showing the location of an accident. If the vehicle goes to the wrong address, by the time it gets to the right place, it may be too late to help. Therefore, public-safety organizations need priority access to spectrum and communication channels.

* Information must be accurate. If a police officer makes an arrest based on erroneous information that a car she stopped is stolen, aside from the considerable distress and inconvenience caused to the person arrested, the city could be liable in a civil suit.

* Security is critical. Public agencies could be subject to penalties if their wireless systems leak unauthorized information. They have a responsibility to monitor and control the release of sensitive information, such as the address of a person with a medical emergency, and to confirm information, such as a report of a terrorist attack.

* These applications are often used in high-pressure situations. Users don't have time to figure out complicated interfaces or to wander around looking for a good connection. They don't want to work their way through multiple screens in order to get critical information. Therefore, mobile applications for this sector must be carefully designed, focused and easy to use.

* Often, workers in this sector use custom devices, either because they have been designed by specialty vendors to communicate with proprietary systems, or because they must be engineered to higher standards than consumer wares. While some organizations have begun to use off-the-shelf PDAs, any wireless application should be able to support non-consumer devices as well.

* Different kinds of organizations have very different requirements, making it difficult for vendors to develop applications with wide utility. For example, the operating requirements for a forestry service ranger would be completely different than

those for a correctional officer who works in the limited area of a prison where it's difficult for cellular signals to penetrate. Some federal agencies need national and even international coverage, while the police force of a small city might be able to rely on a single transmitter for all its mobile communications.

Designing the wireless system and enabling communications is more of a challenge. It's not enough to give public-safety workers access to the data and systems at headquarters. Police, fire and emergency crews pull information from many different locations and agencies. Therefore, there's no single path for the wireless data to travel, and wireless applications must communicate with a mishmash of legacy or proprietary systems software. Interoperability is hampered by the use of multiple frequency bands, incompatible radio equipment and a lack of standardization in transmission formats.

Government and public-safety organizations have one advantage in enabling wireless access to information and applications: Many of them already have a radio communications system in place. Existing buildings and towers can be easily retrofitted to add wireless transmitters. The government even has its own spectrum allocation and a highly developed system for administering and allocating its spectrum.

Government and public agencies should be aware that the technology, hardware and networks are evolving at a rate that can easily outpace their glacial request-for-proposal (RFP) and approval processes. When the RFP, bid and approval cycle lasts years, the implementer may find that the components he's gotten approved are no longer the best choice; they may not even still be available. Many public sector organizations will follow a different path to mobility than businesses, working with technology vendors with whom they have established relationships. Then they can often avoid the need to issue an RFP by describing the mobile application as an upgrade to the existing systems.

Another way government agencies can work around that lengthy approval process is by working with vendors that will bundle the whole solution, including hardware, software, customization, and airtime, and then charge a monthly fee per user. Often this monthly fee is small enough to be paid out of the operating expenses budget. If the mobile application truly cuts costs or increases productivity, going this route could actually improve the agency's operations and free up money for other essential activities.

⊙ WHERE THE BENNIES ARE

Paperwork reduction is one big win, keeping public safety officers out in the field doing their jobs. Because every case is unique, arming field personnel on the run with data such as maps and diagrams can save precious minutes.

Mobile Reporting and Paperwork Reduction

Police departments, fire departments and public-safety organizations all can benefit from reducing paperwork and being able to log information on site while it's fresh. The best opportunities come from automatically logging activities, such as reaching a checkpoint or accounting for personnel. Incident reports written on scene can be transmitted wirelessly or later synched, but this function will only be valuable to officers with good keyboarding skills.

When the organization must report to outside entities or file multiple reports, paperwork reduction becomes less likely, because most government forms are still not provided electronically. Even if they are, it can be difficult to populate multiple forms with data exported from a single mobile application.

Access to Databases

Police departments benefit from access to an array of local and national databases that can help them find out critical information. The right bit of data could help them speed an innocent person on his way or discover that the person they've stopped for speeding is a fugitive. But there's a bottleneck to this plethora of data in the form of the beleaguered operator at headquarters who's responsible for taking information requests, querying the databases and then reporting the results. This system works acceptably most of the time, but can break down when there's a major emergency.

Maps and Plans on Demand

Whether it's a SWAT team attempting to extricate hostages from an office building or a firefighting crew looking for small children in an apartment building filled with smoke, having an accurate picture of the premises can shave life-saving minutes off the operation.

Already, some tactical teams are retrieving maps and floor plans via mobile devices. The South County Fire Protection Authority (SCFPA), a regional public-safety organization in Northern California, uses an application called Premise Information, which lets dispatchers push out such information for "target hazards"; that is, buildings or facilities that the SCFPA has identified as high risk for reasons including a large number of people living there or storage of chemicals. When a fire captain gets an incident address from dispatch, it may have a link to more information that's been entered by the SCFPA, including locations of hydrants recommended for use there and detailed information about chemical hazards on the site.

The next step would be allowing ad hoc access to graphical information as needed. The barriers to accomplishing this are high, because there is no central repository of data. Even different systems within the same entity can't share information. For example, many cities have detailed digital maps of the water and sewer systems, but they reside within a database that's accessible via expensive proprietary software, and other departments that might make use of it don't have access.

Perhaps someday, cities will jettison their crumbling paper ledgers, create digital city maps, and require all new construction to file plans electronically when completed. In order to receive permits for hazardous materials, a business would have to file electronic schematics of its storage facilities, while owners of apartment houses and office buildings would also have to file digital plans. Then, a fire captain could quickly call up the floor plan of a smoking warehouse on his mobile screen to make a highly educated guess about the cause and location of the combustion, or order that a restaurant next door be evacuated.

In fact, a few cities, including Newport Beach, Calif., already maintain such a database of city buildings and provide access to the police and fire departments as well as maintenance crews. (For more information on what the city of Newport Beach has done, see Chapter 9.)

Logistics and Asset Tracking

Whether it's a natural disaster, overseas military operation or a fire, safety and defense teams must mobilize quickly, locating both personnel and equipment, and making sure the right gear goes to the site. A search and rescue team may load 300 boxes of high-tech equipment worth tens of thousands of dollars, including search cameras used to find victims and sensors that determine whether a site has been contaminated with radiation. Each piece of equipment must be loaded correctly when the crew goes out and be easy to find in a situation where a minute's delay can mean the loss of life.

Replacing handwritten packing lists and manual checks with handheld devices equipped with scanners and barcode readers can pay off quickly. Equipment goes on the truck or plane faster when its label is scanned rather than read by a human. On site, boxes can be scanned to quickly identify their contents and where they're needed, freeing up personnel for more critical tasks. Once the team returns to base, logging equipment back in and making replenishment lists is faster and can generate an automatic list which can also tie into the backend systems.

Some logistics systems use Radio Frequency Identification (RFID) tags, small transponders with unique identifiers that are attached to the assets to be tracked. When a transceiver emits a signal over the radio frequency, the tag responds by sending data back. That data could include its location, its identity code or a temperature reading. This data can be used to trigger an action such as opening a gate; it can also create an entry in a database showing, for example that a piece of equipment was unloaded.

The cases that enclose RFID tags can be customized for a particular use. They can be molded into a visitor's badge or permanently welded to an engine block. They transmit their location information to nodes in a wireless local or wide area network, which in turn forwards the information to Internet servers. A variety of wired and wireless devices can access applications that make use of the data.

Personnel Tracking

In the harrowing book and film *Black Hawk Down*, U.S. special forces teams try to rescue two helicopter downed in Mogadishu, Somalia. On the ground, they must run through enemy fire, getting picked off one by one as they desperately try to find their wounded comrades. Meanwhile, their base commanders listen helplessly to radio traffic, unable to provide assistance.

The military's well-established global positioning system would not have helped, even if each soldier carried a GPS unit, because it can tell the user where he is but not his position in relation to others.

There are other problems with GPS: First, it requires a line of sight to a satellite. If the unit is underneath heavy foliage or inside a building, it may not be able to pick up a signal. Second, it's expensive, up to several hundred dollars for a ruggedized receiver, making outfitting an entire battalion costly. Finally, GPS receivers eat batteries; a soldier on patrol for several days would need to slog a pound or more of batteries.

Still, there is a pressing need for technology to track personnel under combat conditions. The military especially fits the profile of an organization that can greatly benefit from wireless applications. Whether in combat or at the home base, its communications model includes a central command center and lots of valuable assets that are often widely dispersed.

The military is testing advanced wireless tracking systems that would consolidate real-time information on the location or movement of assets, whether human or material, and provide it in constantly updated, user-defined reports. These systems will need to provide two-way location information and in many cases won't be able to rely on the GPS satellite system at all.

Use of such systems in combat or rough conditions is probably years away. But several pilots are underway on bases and military training centers.

On the home front, any crew that must keep track of where personnel are deployed, a process known as "personnel accountability," can make use of mobile devices to quickly update the master crew list and, with a wireless connection or via voice communications over radio frequencies, to make sure that the information gets distributed quickly. (For an example of the use of mobile devices for personal accountability, see the case study "Mobile Helps Make Sense of Chaotic Fire Scene" at the end of this chapter.)

Transmitting Medical Information

When paramedics or other medical rescue crews must evaluate and stabilize a trauma victim, every second they save could mean the difference between life and death. They must instantly make critical decisions about the type, severity, and extent of the trauma and, especially when the victim is unconscious or too distraught to communicate effectively, make educated guesses about the victim's general health and possible exacerbating conditions, such as diabetes or heart disease.

If medical technicians could wirelessly request and receive even the most basic medical information about a victim, they could, for example, reduce the chance of giving treatment that might interact badly with medications the victim was taking. The ability to transmit photos and even video could better prepare staff at the emergency room or hospital to deal with the victim's arrival. It would also allow the emergency medical technicians to consult with specialists who could aid in diagnosis and appropriate treatment on the scene.

Video Intelligence and Surveillance

While today's public safety squads use voice over radio or wireless phone to convey information from accident or disaster sites to central dispatch and command centers, in they future they may add video feeds to expand the information gathered. Visual input can help the experienced commanders in the station to evaluate information given by less experienced field officers, let them ask informed questions and help them make better decisions.

Even in more routine operations, a streaming video signal from a patrol car could help station commanders better manage their officers, because they'd be able to offer advice or warnings as situations develop instead of after the fact.

Video streaming wirelessly from a mobile camera could let police detectives be more productive. Instead of wasting time on stakeout hoping that something will happen, they can sit at their desks and do other work while keeping an eye on the video in a small window on the desktop. If a noteworthy event takes place, it's recorded and can be found and saved.

Roaming security guards carrying a wireless handheld device could constantly keep an eye on other parts of the premises as they make their rounds. If a security incident took place in another location, the guard could keep an eye on developments as she was en route there, while security staffers and other key company personnel could monitor the situation and make sure it was resolved.

While the routine use of streaming video is a few years away, at least one demonstration project is up and running. At West Hills High School in Santee, Calif., video cameras mounted on buildings transmit live video to campus security via a fiberoptic network and wireless LAN.

At West Hills, seven campus monitors are responsible for 2,300 students. The 76-acre campus is organized with separate buildings for different departments; classrooms have individual doors to the outside, and there are no interior halls. The unfenced grounds include a canyon which is off-limits during school hours.

PacketVideo, a company that uses proprietary compression algorithms to send video over wireless networks, donated West Hills' video surveillance system, which it calls Skywitness, in August 2002. This "pervasive security system" consists of security cameras, handheld devices and software. (Partners Sony and Cisco Systems donated analog and digital cameras, a local area network and a wireless LAN system.) It's designed for public places, including schools, shopping malls and airports, where it can be both difficult and important to keep an eye on the site from a remote location. Skywitness can send the video signals over wireline, WiFi or wireless networks and can email alerts, video clips and video stills to remote users.

West Hills' security staff watches the live feeds from the security cameras via desktop computers or four IPAQ 3650 wireless handheld devices running PacketVideo's software. Because the system is Internet-based, authenticated users can log in to the application and see the feed from anywhere.

West Hills Safe Schools coordinator Joe Schramm says that while it's too early to quantify benefits such as improved safety and decreased trouble, his gut feeling is that there are "a lot less incidents." He says that knowing they're being watched deters both students and outsiders from misbehaving.

Skywitness also ran tests with the U.S. military to see whether cameras mounted on military ground vehicles could send live video of operations. It even tested the system in commercial airlines, with an eye to giving on-board air marshals better views of the cockpit and cabin. At the same time, Skywitness could stream that cockpit and cabin video to ground control by way of a communications satellite system.

✪ BUILD OR RENT THE NETWORK?

Military and government organizations are in the interesting position of owning their radio networks and, in some cases, controlling some spectrum. This gives them the option of building wireless applications on their own networks. There are a couple advantages to going this route.

First, it can be extremely cost-effective, especially if the community to be covered is small enough to get coverage from a single transmitter site. A wireless transceiver can simply be attached to the city's existing tower. Often, existing IT staff can handle installing the wireless access point at a cost of $20,000 to $50,000. Second, using the proprietary network reduces security concerns that might result from enabling access to the 911 system over the Internet.

Problems

Organizations that can best benefit from using existing proprietary networks are those where Internet access is not necessary, because the force primarily taps into its own computer-aided dispatch system and records.

There are also several disadvantages of building on top of the existing radio system.

* Funding may become a problem, because the network becomes a capital expense and must go through a lengthy approval process.

* Building a proprietary system may limit the department's ability to easily upgrade or take advantage of changing technologies.

* Public entities won't be immune from the same spectrum and coverage that beset all U.S. businesses. Some municipalities might not be able to find suitable radio spectrum to license, or the available channels may be too slow, especially if the force hopes to use data-heavy applications such as downloading photos and maps. Working with a commercial carrier insulates the organization from worrying about network upgrades and spectrum.

* Using existing communication channels will deprive public-safety organizations of what could be a prime benefit of wireless: the ability to quickly share data across organizations; for exam-

ple, during a highway crash where fire trucks, ambulances and police cars converge to do their parts.

The Public Safety Wireless Advisory Committee (PSWAC) was established by the Federal Communications Commission and the National Telecommunications and Information Administration to study the spectrum needs of public safety organizations and create a plan to ensure that those needs were met. On September 11, 1996, the PSWAC filed its report with the FCC.

The PSWAC stated the problem with chilling accuracy. It said: "In an era where technology can bring news, current events, and entertainment such as the Olympics to the farthest reaches of the world, many police officers, firefighters, and emergency medical service personnel working in the same city cannot communicate with each other. Congested and fragmented spectral resources, inadequate funding for technology upgrades, and a wide variety of governmental and institutional obstacles result in a critical situation which, if not addressed expeditiously, will ultimately compromise the ability of Public Safety officials to protect life and property."

According to the PSWAC Steering Committee: "The availability of commercial systems as a reasonable alternative [to proprietary public safety networks] depends on satisfying several essential requirements. These are: 1) high reliability coverage throughout the area designated by the agency; 2) affordable cost; 3) priority access during peak periods and crisis circumstances; 4) secure transmission including, in particular cases, encryption; 5) sufficient reserve capacity; 6) reliability comparable to dedicated systems; and 7) mobile and portable units distinguished by the durability and ergonomic factors required by field personnel."

Table 8.1 Ten-Year Life Cycle Costs—Wireless Data System

Agency Costs	Purchase	Service Agreement
Infrastructure cost	$500,000	$0
Mobile equipment (25 cars @ $5000 each)	$125,000	$125,000
Mobile software (25 cars @ $800 each)	$20,000	$20,000
Airtime (25 cars @ $50/month/car)	$0	$150,000
Infrastructure maintenance (approx. 10% of purchase price/year)	$500,000	$0
TOTAL:	$1,145,000	$295,000

✪ THE PUBLIC SECTOR CHALLENGES

Although many of their tasks are similar to those of private field forces, public organizations have some unique challenges.

Security

Listening to police radio transmissions using a scanner is a popular spectator sport. But police departments routinely transmit data that's highly sensitive in terms of public safety and individual privacy, so their data transmissions must be bullet-proof. The stakes are higher—life or death—for soldiers in the battlefield. For many public-sector organizations, security decisions take precedence over all others.

The first task in planning and implementing security for a military or public safety organization's wireless data applications—and this is just as true for businesses—is to set a policy for the level of security necessary for the various data communications. As part of this process, the decision makers should identify the special vulnerabilities of the different connectivity strategies they're considering.

Next, they should map these various policies to published standards, such as the Federal Information Processing Standards for information security. Once clear policies and standards are in place, it becomes much easier to take the third step and evaluate the technology products that are available to secure communications.

In making these decisions, it's important to remember that every security measure involves some inconvenience or delay for end users, whether it's requiring them to type in a password or to install an extra piece of hardware.

Security strategies include the following:

* **Authentication:** The system can be designed to require users to log in with a user ID and password, or to insert a physical key or smart card that handles authentication. Biometric authentication uses physical characteristics of the person, such as the pattern of the retina, a fingerprint or a whole handprint.

* **Encryption:** Data can be encrypted using client/server software applications, or by physical devices located somewhere between the device and the servers it accesses.

Many wireless applications have built-in security; these or off-the-shelf security applications can be a good start for lower-level security needs. Wireless networks should be thoroughly and consistently tested to make sure they're secure.

The San Mateo, Calif., Police Department added Media Access Control (MAC) address filtering to its 802.11b wireless network. The MAC address is a unique identifier for a piece of hardware. To make this work, IT administrators must maintain a database containing the unique MAC address of each device that is allowed on the network. The wireless access point checks the MAC address of a device requesting access, and if it's not on the list of known good MAC addresses, it refuses the connection.

Virtual private networks (VPNs) can help secure transmissions. The owner of a particular database—the state Department of Motor Vehicles, for example—can set policies for access to its VPN and rules for what parts of its driver's license information can travel over it. For example, the State of California Department of Justice (DOJ) doesn't allow an individual's criminal history to be transmitted to mobile terminals. At the same time, the California DOJ dictates technical standards for the networks on which its database information travels, including a requirement for use of an encrypted VPN.

The San Mateo Police Department also compartmentalizes its wireless traffic, both for security and also for speed considerations. Text-based information, such as queries to state, local and federal databases, go over the proprietary radio network. The department has begun transmitting photos; small thumbnails are sent wirelessly to the police car laptop using the radio network, while officers drive to an 802.11b node to download larger photos over the broadband connection.

While police departments may not need to secure their radio dispatch communications, as they move more of their reporting and record-keeping out of the office and into the field, logging incident reports and writing narratives on mobile devices, they must be able to prove that the information generated hasn't been tampered with or compromised. For example, they'll eventually be called upon to prove that the description of what happened during an arrest hasn't been altered after the fact or been accessed by an unauthorized party, such as a newspaper reporter or attorney.

The FUD factor

Public sector technologists may have to deal with the FUD factor when introducing wireless apps: the fear, uncertainty and doubt that often accompany a change in work procedures. As with general field-force applications, every public safety officer won't adopt wireless applications with the same enthusiasm as their buddies in the IT department. In some cases, it comes down to personal preference and hard-dying habits. Personnel of a certain age may not have learned keyboarding skills in school; they'll find it faster and less frustrating to stick with the pencil.

Public-sector organizations may find it difficult to work with technology vendors used to dealing with commercial companies— and vice versa. Government organizations have different priorities and business needs. Commercial vendors can quickly get frustrated with the long sales cycle and the municipality's constantly changing financial wherewithal, and decide their sales force's time can more productively be used selling to private sector customers.

It's Babel Out There

Because they work with so many outside agencies, many of which are not very automated, it may be difficult for public entities to get the same benefits in paperwork reduction as do private-sector field forces. The police department originates paperwork that will be used throughout the criminal process. There's a little room in the South San Francisco police station, for example, that has an entire wall of paper report forms: forms for the California Highway Patrol, forms to send to the Department of Justice following an arrest, forms for reporting the possible use of a car in a crime to the Department of Motor Vehicles, forms to help the District Attorney's office decide whether to prosecute a case in court.

While South San Francisco police officers have mobile access in their cars, which can access the central dispatch computers and state and local databases, they still have to spend as much as three hours of their shift in that little room, filling out reports by hand. Filing reports from the car would be more efficient and also make them fresher, but the department hasn't found a way to digitize all those different forms, nor to be able to export digital information to the many different outside systems that need it.

Things get more complicated with organizations that work on a national or international scale. The Department of Defense (DOD) cooperates internationally with a variety of allies and organizations employing different cellular technologies and communication protocols. Many of them rely on legacy systems different from those of the DOD. After doing the integration work necessary to enable allied organizations to share mobile data services, it's quite a challenge to later upgrade or change them.

It's unfortunate that it's taken tragedy to awaken an understanding of the pressing need for fast and unified access to critical information in the public sector. With a new sense of urgency, public organizations are racing to identify and implement wireless and mobile applications in a way that may transform their own operations. And, just as the military's DARPANet evolved into the commercial Internet, the decisions made by the national and local governments may transform aspects of business and society in ways that can't yet be imagined.

Making Interoperability Work

By Tom Hoyne

For first responders such as police, fire and emergency personnel, it's not a question of *if* the next large-scale crisis will happen, but *when*. Will they be prepared? Events such as 9/11—large-scale emergencies that require coordinated communication between multiple first-response agencies—present a scenario in which public-safety officials lack the ability to communicate effectively to conduct operations and save lives. This lack of communication, well-documented in reports and investigations following the attacks, illustrates a serious problem that the public-safety community has faced for some time.

Across the country, public-safety agencies lack interoperability—the ability for disparate radio systems to communicate when needed. Prior to 9/11, many public-safety wireless advisory groups, such as the Public Safety Wireless Network (PSWN), Federal Law Enforcement Wireless Users Group (FLEWUG), National Coordination Committee (NCC), and Association for Public-Safety Communications Officials (APCO) identified interoperability as one of the most critical issues faced by public-safety and security organizations, and began working to address it. Heightened interest in homeland security and emergency-response preparedness after 9/11 gave the public-safety community the opportunity to participate in discussions about interoperability at the federal level.

NATIONAL INTEREST
FOR INTEROPERABILITY

In November 2002, Congress approved the Homeland Security Bill, realigning several federal agencies and providing new funding for emergency preparedness. One of the key initiatives described in the

National Strategy for Homeland Security is to make existing first-responder communication systems interoperable for quick and effective coordination of necessary response assets. Mike Byrne from the Office of Homeland Security has said, "My idea of music from God is interoperability."

Congress has recommended that 2002 Supplemental and 2003 Budget funds be made available to states for network-level solutions that make existing communications interoperable. Under the auspices of the Federal Emergency Management Administration (FEMA)'s proposed legislation, grants will be available for each state to purchase cost-effective solutions such as cross-band repeaters, frequency band patches and other network-level solutions.

Now, the question is: What is the best solution and what are the considerations for the states tasked with developing an interoperability plan?

CHALLENGES OF INTEROPERABILITY

As state governors review their interoperability needs and strategy, they are evaluating several factors. How many potential users are in the state? What is the cost per connected user for the interoperability solution? Does it have the capacity to not only connect my state but also connect my state to others in my region? Who can talk now who couldn't before? Is anyone left out? Do state agencies have to buy new radios? Will new equipment be compatible with the old? Does the equipment work with other vendors' equipment?

One only needs to look at the various frequencies used by local, state and federal public safety agencies to understand the difficulty in finding an appropriate interoperability solution. The various agencies operate in different frequency bands using different protocols and different equipment. Systems include VHF, UHF, 800 MHz, 900 MHz, trunked and untrunked, and digital and analog equipment. Take into account that equipment from different vendors is not compatible, and the solution is even more muddled.

In choosing the best interoperability solution, agencies must not only consider cost of new equipment but also amount of time needed for system deployment, investments in training for personnel to become proficient with new equipment and ability to support future growth needs. All of these factors are significant in evaluating the best solution for interoperability.

Cost

For the public-safety community, system and equipment costs are an ongoing worry. Land mobile radio systems for public-safety communications are significant investments for local, state and federal agencies. As a result, the cost versus benefit of new investments must be carefully weighed. Even with federal funding, each state must determine where to allocate money to best meet its interoperability needs. The decision becomes how effectively can existing investments be utilized while still providing the capability and features necessary for effective operations? At what point is it a better option to buy entirely new systems or radios?

System Migration and Deployment

For public-safety agencies, technology change is a significant concern. The availability of new, more-efficient wireless modulation schemes can enable users to get more out of their limited spectrum resources. Ensuring that new systems are compatible with legacy systems is an important factor for towns and counties that wish to swap out equipment over a period of time. First responders often prefer to use existing radios until their own needs change. As a result, new technology must be compatible with older technology but allow for future upgrade.

Migrating a state-wide radio system also takes time. Typically, installing a digital system to replace an analog system takes three to five years before the system is fully functional. The major factors are time associated with land acquisition, construction and training first responders on new equipment. For multiple states or a national system, deployment time would take much longer.

Frequency Availability

About 10 different frequency bands exist for public safety. A true interoperability solution must have the ability to be cross-band; for example, enabling VHF radios to talk to 800 MHz radios. An effective solution cannot be reliant on frequency availability to accomplish cross-banding between different radios.

SOLVING THE INTEROPERABILITY PROBLEM

The need for coordinated public safety communications through interoperability is clear but how do state agencies, local departments and neighboring counties, which each buy separate equipment, achieve this end? Decision makers have three options—buy new radios, install hardwired or software patches, or link existing systems with a network-based solution.

A Radio-Based Approach

The traditional suggestion has been to move all first responder departments and agencies to a single radio-based system, requiring state and local governments to replace all radio equipment and infrastructure with new standards-based equipment. In its simplest terms, this approach to large-scale radio-based interoperability would require all public-safety officials to use identical radio hardware.

However, due to the tremendous costs, time needed to install new equipment and frequency limitations, this approach is not feasible. PSWN estimates the cost of building such a public-safety wireless infrastructure for the U.S. at more than $18 billion. As the recent Homeland Security Bill left out funding originally promised for local first responders, the cost of achieving interoperability with all new equipment is unrealistic for any agency. Most first responders continue to use analog technology because they cannot afford the latest digital systems. In fact, these agencies account for the majority of public-safety radio systems in the field today.

Even if full funding was available, the entire implementation would take as long as 10 years to implement without any certainty that there would be enough capacity on shared frequencies to meet large-scale needs. An effective radio-based solution relies on an abundance of frequencies that are not available.

Hardware and Software Patches

A second option for interoperability is use of hardware or software patches that link talk groups in response to onsite situations. This equipment has the advantage of enabling interoperability using current radio systems. It provides a cost-effective solution for average day-to-day communications needs in the short term. However, as a long-term solution to large-scale interoperability, patches have several drawbacks.

While excellent for enabling a situational, hardwired connection between talk groups, patches are not dynamic and are therefore unable to easily accommodate unplanned participants. The ability to regroup and change communications groups in an emergency is one of the most important requirements of effective interoperability. True interoperability does not require human intervention, whereas patches require dispatchers to reconfigure interoperability structures.

For large-scale interoperability, patches also lack the capacity to handle tremendous call volumes. Channels are shared and may unexpectedly not be available for communication. The limited regional reach of patches does not address the interoperability needs for a large, multi-agency response.

A Network-Based Solution

A third alternative is to achieve large-scale interoperability *via* a network-based solution, using principles similar to those of the communications networks used by Internet providers and voice carriers. At the core of wireless public-safety systems is a wireline backhaul network. Therefore, the key to wireless system interoperability is to connect systems at the network level. Such a network would link existing systems into regional, statewide or national systems to create a private intranet for multi-agency interoperability, without requiring the purchase of new end-user equipment.

Application of Internet protocol (IP) as a communication standard for network connectivity between systems offers a cost-effective and logical solution. Developed during the 1980s, before the explosion of Internet and packet-switch technologies, public-safety wireless networks for critical communications lack the benefit or heritage of IP. Traditional land mobile radio (LMR) systems used by fire departments, police and other first-response agencies began life in a world of circuit-switched network backhauls, proprietary protocols and custom switching hardware.

Today, the ubiquity of IP technology makes the installation of IP-based network connectivity an easy and natural progression as hardware and software evolve. Because IP-based wireless systems can be assembled using off-the-shelf components, the cost is relatively low for providing the interoperability, high performance and reliability that public-safety systems require. In fact, building such a network-based interoperability solution would cost around $1 billion, a fraction of the cost of a radio-based solution. More impor-

tant, network-based interoperability does not require purchase of
new radios, involves significantly less infrastructure investment,
provides easy compatibility with disparate systems and quick time
to implementation, and offers flexibility for future upgrades.

USE OF IP

The public-safety industry's need for 100 percent reliability, wide
area coverage, and interoperability using limited frequency bandwidth and disparate legacy systems can be met with application of
IP. It also offers the capability for future enhanced wireless data
delivery to make public safety workers more effective.

The specific requirements of critical communications networks—
reliability, fast access, multiple users and agency broadcasts, and
transmission of high-quality voice and data—all can be met using
a similar architecture. A well-designed and -managed private IP-
based intranet can be tailored to provide the fast response times,
excess network capacity, high-quality voice and security, and interoperability that are essential to public safety. IP provides a common standard for various wireless systems to communicate using
multiple brands of radios and operating seamlessly under the most
demanding situations.

An IP network-based solution also has the advantage of sharing an IP backbone between the agencies, which allows virtually
limitless expansion of the network, without costly full system
upgrades. There is also no need to dedicate resources to training
first responders. Because existing equipment connects at the network infrastructure level, communications appear seamless to the
end users. First responders can upgrade individual radio systems
over time as departments desire and as budgets allow and connect
back to the network.

INTEROPERABILITY NOW AND BEYOND

Recent tragedies have forced federal, state and local governments to consider achievement of direct and immediate interoperability between public-safety personnel as a major priority. Emergency-response preparedness is essential, since there is no way to predict when another large-scale crisis will occur. An IP-based network solution can ensure that agencies have the level of interoperability on demand that they need to coordinate a large-scale emergency response.

Network-based interoperability can connect all systems, with the flexibility and scalability to accommodate the various needs and challenges that public-safety officials encounter. More important, it provides each first-responder agency with the ability to use current equipment and upgrade at its own speed as its needs change, while maintaining the interoperability capability required on a national level.

Tom Hoyne is manager of marketing for wireless systems for M/A-COM, a developer and manufacturer of radiofrequency and microwave semiconductors, components and IP networks. A graduate of Purdue University, Hoyne has been involved in sales, product management and marketing of land mobile radio networks at M/A-COM's Lynchburg facility for 10 years. His prior experience was with General Electric, marketing utility and industrial products throughout North America.

CASE STUDY:
Mobile Helps Make Sense of Chaotic Fire Scene

Fighting a fire is just like building an enterprise, says fire battalion chief Richard Price of the South County Fire Protection Authority (SCFPA). The difference is, you need to build the whole organization in minutes. "As you arrive at a fire," Price says, "you put some bodies in charge of attacking the fire, some in charge of making sure adjacent buildings don't catch on fire, somebody in charge of safety. Meanwhile, I'm trying to build an organizational hierarchy on the fly."

Things get more chaotic as the firefighting progresses. There are people arriving from different directions and getting to work on attacking the fire. Some units may have arrived before the commander and have gone to sides of the building that the incident commander can't see. "Let's say 15 minutes into the incident, the second floor collapses. You need to know, 'Did everybody get out?'" To figure it out, Price needs to know not just who was dispatched, but who is physically there.

It gets harder still when the incident is something like a wildfire that may go on for several days. Then, you have to feed the crew, provide bathrooms and a place for exhausted crews to nap, and a way to pay for consumables like food and fuel. You might have to coordinate air support or "con crews" that come from prison to help out. Tires get blown, uniforms get ripped, crew members get injured. It's a logistics nightmare.

Under pressure-cooker conditions where a missing slot in the org chart could lead to overlooking personnel trapped in the burning building, battalion chiefs like Price can't afford mistakes. As crews and equipment arrive, the incident commander must build a hierarchy of personnel that's organized much like a business, with different units spreading out under supervisors who report to the chief. Yet, despite the deadly urgency, the majority of commanders keep track of this constantly shifting economy with hatch marks on a yellow pad or, at best, with an erasable white board in the command truck.

Not Price. Since 1999, the SCFPA, which provides fire and emergency medical services to three towns in San Mateo County, Calif., has provided its staff with off-the-shelf wireless Palm devices running FireDispatch, a suite of 14 applications specially developed for firefighters. On the way to a fire, Price uses his Palm to log on to Personnel Accountability System (PAS), a FireDispatch app that lets him see what engines and trucks have been dispatched, what firefighters are coming with the trucks, and what their expertise is. Does he have a hazardous materials specialist? How many emergency medical technicians will arrive?

When he gets to the scene, Price begins to build his org chart. As trucks arrive at the fire, the driver pushes a button that automatically signals the CAD system. When Price refreshes his PAS screen, he can get a list of all firefighters who are physically on scene; retrieving this information usually takes just a couple seconds. With voice radio, Price takes roll call using this list and, as units respond, he checks them off onscreen. He can also note squads' locations or assignments. "Now, as the units come in they can see who's doing what, what sort of organization has been built so far." When the roll call is complete, pressing a screen button sends the information back to the dispatch server at the Public Safety Answering Point (PSAP), the unit that handles 911 emergency calls. The computer updates every other user's screen and also enters the information into the incident log.

Generating automatic log reports like this helps the commanding officer document that staff are following accountability procedures, which require that the incident commander check in with all crews every 30 minutes.

FireDispatch runs on almost any mobile device, including Black-Berry two-way pagers and WAP phones. It directly connects with the computer-aided dispatch software used at PSAPs. SCFPA chose the Palm platform for several reasons: First, the device and inter-face were easy to use. "We have a wide variety of user skills and com-fort levels with technology," Price says. "When you try to introduce new technology, you always have to provide it for the lowest com-mon denominator." Second, the Palm operating system makes it easy to download and install applications; this eases the burden on the IT staff. Firefighters rely on Palm's tech support staff for most questions or problems.

"You can go to Office Depot," Price says, "buy a Palm, down-load the apps, synch your device, register with the dispatch center electronically and be up and running in one day."

Price likes FireDispatch because the various applications are so simple to use. "It's not an application that does 10 things," he says. "Instead, we have lots of applications, but they are all very focused. They do one or two things very well." He *should* like it. His depart-ment helped refine and test the software, which was designed by Telecommunications Engineering Associates, a software company headed by Daryl Jones, a former police officer.

Because of the critical nature of their work, SCFPA firefighters can't wait for a real fire to practice using the application. The Fire-Dispatch Web site has demos of the different applications, so that users can practice going through the steps without having to waste time or get distracted during a real incident.

While the rank and file has standardized on the Palm i705, fire chiefs, inspectors and commanders are allowed to choose the device they'll use. Neither is FireDispatch a mandatory application, though it's seen as a standard way of doing business. "If you've been a fire chief for 20 years, and you have three years left before you retire," Price says, "it's not so easy to get people to switch what they do."

THE NUTS AND BOLTS

Hardware: Palm i705, various laptops and WAP phones

Applications: FireDispatch

Number of users: 100

Implementation time: One year

Cost of project: Around $500 per user

Network service provider: Palm.Net, AT&T Wireless CDPD network, commercial cellular (depending on the device)

VARs/application vendors: Telecommunications Engineering Associates

CASE STUDY:
Shipboard Inspectors Carry PDAs
From Bow to Stern

The Navy's Military Sealift Command provides a vital link between U.S. forces worldwide and the portside resources they need when they're standing by, ready for combat.

MSC's 35 Naval Fleet Auxiliary Force (NFAF) includes oilers, ocean tugs, fast combat support ships, combat stores ships and ammunition ships. They supply food, fuel, ammunition, spare parts and other supplies to Navy combatants on any ocean around the world so they don't need to return to port. Two hospital ships stand by on the U.S. coasts in reduced operating status to provide emergency medical care for combat troops if the need arises.

The 37 MSC Prepositioning Program ships strategically place combat equipment, supplies and fuel for the U.S. Army, Marine Corps, Air Force, Navy and Defense Logistics Agency in three locations near the world's hot spots, ready to supply U.S. forces for any contingency. The ships include long-term chartered commercial vessels, activated Ready Reserve Force ships from the U.S. Maritime Administration, and other U.S.-government-owned ships, all crewed by contract mariners.

To make sure that MSC's ships remain shipshape, teams of MSC inspectors perform evaluations every six months, opening up and checking all major equipment—including the engineering plant, the steering system and the record-keeping systems—to get a feel for the readiness of the vessel.

Teams of 25 to 30 inspectors assess NFAF ships using a system known as Ship Material Assessment and Readiness Testing (SMART), while teams of two Contract Quality Assurance (CQA) inspectors oversee the contract operators of the Prepositioning Program ships.

"These are 1000-foot ships," says Military Sealift Command project manager Mark Andress, "that create their own propulsion, generate their own electricity and run their own air conditioning. They're very complicated."

The inspectors comb the big ships, starting at the exhaust stacks and finishing in the bilge. It's rough work, with inspectors descending on ladders into holds, treading on catwalks and crawling behind machinery. The inspectors document deficiencies in the ship's condition and maintenance program, and at the end of each day, the inspectors deliver a report to the ship's maintenance engineer.

At the end of the inspection, the ship gets a readiness score, and the list of outstanding deficiencies is given to the ship's chief repair officer, who can either turn them into work list items for the ship's crew or send them to the shore engineer to get bids for commercial repairs.

When this operation was done using the pencil-and-clipboard method, inspectors would crawl the ships each day, then spend all night producing the deficiency reports from their paper notes. "At the end of the inspection," Andress says, "they'd be wiped."

There were other problems. It was easy to lose track of deficiency items. There was no way to make sure that the ship's repair officer had received the entire report. And sometimes headquarters wouldn't see the results for months, because the ships were so far-flung.

In the early 1990s, MSC replaced its pencil-and-clipboard inspection method with the SMART system, a software application created by Seaworthy Systems, an integrator specializing in marine engineering, that let the inspectors document deficiencies on laptop computers. However, laptops didn't seem like the best solution for automated entries. The inspectors didn't like to carry them around, because they were afraid they'd drop and break them. So mostly the inspectors continued to make their handwritten notes, then circle back to the central location where the laptops were set up and type the notes into the laptops.

In 2000, MSC turned to mobile applications vendor iAnywhere to port the SMART application to mobile devices. The idea was that the SMART inspectors would keep their handheld devices with them as they crawled and climbed through the ships, then synch the information to the laptops. However, the SMART inspectors didn't see enough advantage to using the handhelds. MSC also tested using a wireless connection to transmit the information from the mobile devices back to the laptop hub, but the huge metal ships make radio connections difficult. Neither was there much advantage to wireless to offset the additional hardware costs. Turning in the deficiency reports at the end of the day worked just fine. The SMART teams stuck with their paper forms and laptops.

However, in 2001, the CQA inspectors heard about the mobile SMART and, as Andress says, "picked it up and ran with it." The MSC information technology team reworked the mobile SMART application to suit the CQA inspectors, calling it Snapshot. Now, as CQA inspectors move around the ship, they can log deficiencies on the spot. The interface has checklists with standard deficiency items such as "rust" or "loose bolt." Inspectors can also write in items or notes, building the deficiency report as they work. At the end of the day, they simply synch their devices to a laptop.

The CQA inspectors can now complete the inspection of an entire ship in one eight-hour shift instead of two long and exhausting days. "Now," Andress says, "they leave the ship at 6 PM, and they're done. And now, when they come to the office they're working on other issues, not just retyping their notes [into the desktop computer]."

NUTS AND BOLTS (CQA)

Hardware: Palm 505s

Applications: Snapshot, iAnywhere's SQL Anywhere Studio

Number of users: Eight CQA inspectors

Implementation time: Four months for initial release of SMART, two months to modify it for Snapshot

Cost of project: CQA Team hardware (two laptops, two Palms) $5,000 to $6,000 per two-person team; software costs not disclosed

Network service provider: None

VARs/application vendors: Seaworthy Systems, iAnywhere Solutions

---

CASE STUDY:
Mobile Gets Rescue Team to Disaster Sites Fast

Whether it's a natural disaster or man-made, protecting U.S. citizens on their home soil is the duty of the Virginia Task Force 2 (VATF-2), one of 26 teams designated by the Federal Emergency Management Agency to respond to urban disasters or attacks. VATF-2 rushed to Washington, D.C., to help in the September 11 Pentagon rescue efforts. Trained to deal with weapons of mass destruction such as anthrax, at full strength and hosted by the Virginia Beach Urban Search and Rescue Team, the team has 70 members—engineers, doctors, heavy riggers, canine search, civilians and fire fighters—who respond to disasters.

To mobilize quickly, VATF-2 must load as many as 300 heavy plastic transport boxes, each holding from 2 to 300 different items, any one of which might contain some critical piece of rescue equipment. The boxes are loaded onto military pallets that are about 60 square feet; some of them are stacked eight feet tall with boxes. It's important to keep track of how they're loaded so the team can quickly find what it needs when it hits the field.

VATF-2 used to rely on typewritten lists attached to each box to tell what assets were inside and to check off returned pieces of equipment. That was hard enough when shipping out to a disaster site; it could be impossible in the chaos of the base of operations (BOO).

In the early 1990s, VATF-2 decided to computerize its asset tracking system. "When you have 15 or 20 people wanting items," says VATF-2 logistics manager Wayne Black, "you can't stand around looking through your list."

A programmer associated with the team developed a computer-based application called TAVALSS (Total Asset Visibility And Logistics Support System). TAVALSS automatically fills in forms show-ing where assets were stored. The team used barcode scanners to get information off the boxes and into the computer.

In 1994, the team added handheld scanners from Intermec that could transmit the scanned information to the host computer using radio frequency (RF). Then, in 2002, it switched to PocketPC devices with integral scanners, also built by Intermec, because the com-pany stopped making the original model. These handhelds were equipped with wireless network cards; they transmit information to the host using an 802.11b network. The devices still report to a central computer that propagates the data to all the other devices on the network, so they are all constantly in synch. This method works better under field conditions, where the relatively limited range of the 802.11 network (about 1,500 feet) is plenty for a BOO.

The new application has several modules besides Cache, the logistics tracking app; the other applications manage information on personnel, deployments and equipment check-outs. It also allows for different access levels in the application. For example, doctors on the team can view personnel medical records unavailable to other members.

"Now, we can palletize our entire gear in less than an hour," Black says. "It used to take us two or three hours when we used paper. At the Pentagon, we moved people and their gear in and out in 10 or 15 minutes."

The handheld scanners help if the barcode gets ripped off the box. Each piece of equipment has a number stamped onto it; the crew can scan that number and query the application to find out what box it should be put in. The team also uses a third-party form generator that ties into the TAVALSS system.

Black says the new system lets VATF-2 go home with almost all the gear it brought, a vast improvement from the paper and pencil technology. It's also freed the team from having to do constant full inventories of the equipment cache when it gets back to Virginia Beach. "For the most part," Black says, "when we leave there we've already taken each item and put it back in its proper location."

NUTS AND BOLTS

Hardware: Intermec ruggedized PocketPC

Applications: TAVALSS

Number of users: 25

Implementation time: Six months

Cost of project: Not available

Network service provider: Private local network

VARs/vendors: Intermec

Chapter 9
Telematics Drives Business

Telematics as a concept is often met with some confusion; many people think of telematics in terms of consumer-oriented systems, such as GM's in-car OnStar product, which provides drivers with navigation and information services. The technical definition of telematics is the integration of telecommunications and computing, but the term has come to be used to describe Internet-connected computing systems in vehicles. The real-world enterprise telematics market makes this definition much clearer.

There are the more obvious telematics verticals, such as trucking and shipping. Fleet management is definitely a top application—Allied Business Intelligence estimates that the market for telematics-based commercial fleet management systems is expected to grow from under $2 billion in 2001 to nearly $6 billion by year-end 2007. But there are also some very interesting and lesser-known applications, such as leased-equipment tracking and agricultural yield management. What all of the markets have in common is that their primary work force is mobile within vehicles for much of their day, which presents some unique workflow management, safety and customer-service issues.

For many trucking companies, the savings on insurance premiums alone make a telematics system worthwhile. With post–September 11 insurance-rate increases, even small-fleet owners are now turning to telematics to lower rates, increase security and better manage a mobile workforce. Accordingly, the price of deploying a telematics solution is dropping, making the decision easier for smaller companies.

Though it's helpful for many companies to have access to the wealth of data that telematics systems can collect (for analysis and forecasting), it's usually the data that reflects exceptions to the norm that are most helpful in the day-to-day work environment. Vendors are calling this "exception-based data management," and it's the way of the future for telematics. For example, instead of the telematics system telling a dispatcher every time a certain truck moves within a few blocks, perhaps it's more helpful for that dispatcher to know only when the truck leaves a certain geographic area. In this case, a company would set up a "business rule" that said: If truck X moves out of predetermined geographic area Y, then send an alert to manager Z. If manager Z doesn't respond within five minutes, then send an alert to manager Z's voice mail, or to manager W instead.

⊛ TELEMATICS BRINGS EFFICIENCY GAINS TO WORKERS ON THE GO

Beyond the predictable market of fleet and delivery vehicles, there are a number of industries that can benefit from telematics applications. As services become more widely available and more affordable, smaller markets are finding value in outfitting segments of their workers with telematics tools.

Email/Information Services

Whether you're dealing with truckers who spend just about all of their working lives in their vehicles or delivery drivers who use their trucks as surrogate offices, wireless systems can be used to transmit critical information to workers on the road. For example, schedule and delivery updates can optimize a delivery driver's time on the route. Weather and traffic updates can do the same. And the ability to read and respond to email can make life on the road or the sea a lot easier—and more productive—for truckers or shipping crews.

GPS Navigation

Navigation via global positioning system (GPS) gives drivers precise directions and fleet managers the exact locations of their trucks. Any business that has vehicles in motion can use a GPS system to relay precise directions to its drivers. Real-time, turn-by-turn GPS navigation is especially useful for emergency vehicles that need to get to an incident quickly.

Fleet Management and Inventory Tracking

Fleet management is definitely the predominant telematics application. Applications include everything from tracking truck locations to conveying a wealth of data about vehicles—diagnostics, fuel levels, cargo weight. Inventory may also be tracked independently of trucks to ensure delivery and maintain security.

Customer Service

In many cases, the ability to offer superior customer service is the number one benefit of a telematics application. This is especially true for trucking companies with customers that are dependent on timely shipments. The ability to track cargo and trailers on the road lets trucking companies give customers up-to-the-minute updates.

This can be an important differentiating factor when a customer is choosing which trucking company to go with. The use of bar-codes and radio frequency identification (RFID) tags lets shipping and trucking companies give their customers the ability to track specific items within a truck or shipping vessel and confirm location via Web-based tracking programs.

Mobile Workforce Management

Companies with field-force workers can make better use of workers' time by tracking location and job status. If one worker finishes a job early, she can be dispatched to the next site and other workers' schedules can by adjusted accordingly in real time. A system that provides managers with the ability to coordinate their teams this way acts as a "proxy"—it takes the place of having a real manager out in the field with workers. J.D. Fay, vice president of corporate affairs at wireless service provider @Road, calls this "mobile resource management"—the automated allocation and use of mobile workers and assets while they're out in the field.

For companies that see new customer lists every day, having a wireless system that can manage changes can save a lot of manual labor. Real-time routing allows companies with service technicians in the field to make appointments that are much more convenient for their customers—a customer with a bad DSL line might be able to get a service technician who's in the neighborhood, rather than waiting a few days for a scheduled appointment. The major benefit for companies using wireless systems with their field-force workers is not always the day-to-day efficiency improvements, but rather the data available for analysis—how many jobs a worker is able to do each day, how many miles a worker drives each day, how much time is spent in the car versus on the job, how many workers are in a particular geographic region. Savings come in the form of companies completing more jobs per day and seeing less downtime and "windshield time" from field workers.

Streamlined Business Processes

A good telematics system that integrates with a company's existing payroll and invoicing system can save time and money. By automating compliance services such as fuel taxes, companies save the paperwork and labor that normally goes into preparing documents. Trucking companies that have automated the dispatch and routing process may be able to add more vehicles to their fleets without adding more dispatchers, which adds up to more profits in the long run.

✪ A VARIETY OF MARKETS BENEFIT FROM TELEMATICS APPS

It's easy to imagine the benefits of simply giving mobile workers the ability to perform computing while on the go. But the real payoff in the telematics market comes from specific services and applications designed to suit the needs of specialized groups of mobile workers.

Fleet Owners

This market can be divided into two segments—large-fleet and small-fleet owners. Large-fleet owners, such as trucking companies that own thousands of flatbeds and coordinate travel all over the United States, have been using telematics applications for years. Because telematics has historically involved a steep hardware and services investment, smaller-fleet owners (a delivery service with 50 trucks in one region, for example) have found it difficult to justify the cost. But the market is becoming more and more accommodating for smaller fleets, and so short-haul trucking companies are now entering the telematics market.

One of the biggest benefits for fleet owners is the automation of business processes. "We're really putting an office in a truck," says Bob Montana, president of telematics provider Summary Systems. Telematics systems can reduce a tremendous amount of paperwork—dispatches can be sent directly to trucks and delivery information can be sent directly to the main office. Companies can just download data off trucks at end of day, rather than transcribing the driver's notes. Companies can pay their highway usage taxes wirelessly, rather than paying a bookkeeper to go through records manually.

Controlling drivers and delivery schedules also becomes less time-consuming with an automated process. In many cases, telematics systems for fleet owners don't have to be real-time—they can be synch-based and data can be downloaded at the end of day. Especially for small-fleet owners (fewer than 50 trucks), business processes such as billing, repairs and load management can be overwhelming. A good telematics system can automate a lot of these functions, which is key for smaller operations that often don't have a large back office staff to manage business functions and handle paperwork.

Customer support is also a driving force behind telematics in the fleet management sector—companies can give a customer accurate information about a delivery (without having to call every truck stop in the U.S.).

From the driver's perspective, there's more access to help if there's an accident or breakdown. Ron Konezny, chief financial officer of telematics provider PeopleNet, says there have been about a dozen situations where a wireless data communications system has saved the life of a trucker with a health problem. A driver who's having a heart attack can alert the dispatch center and a GPS system can locate his or her truck quickly for the paramedics. However, drivers have also had negative attitudes to telematics systems, as some feel they're being "tracked" along with the trucks. So, educating employees and involving them in discussions about the benefits of telematics systems is key to user adoption.

Taxis and Limousines

Like any other business with fleets of trucks or cars, taxi and limousine services are using telematics systems to improve dispatch services—routing taxis according to their locations in relation to incoming calls saves companies and drivers time. Billing systems and ordering systems can be wirelessly integrated to create less work for the corporate back office. And more data for corporate analysis is always a bonus as well. Allied Business Intelligence analyst Frank Viquez says he's seeing this market grow as a lot of taxi and car service companies consolidate.

Driver safety is also a driving force in the taxi industry—vendors are offering telematics systems that give drivers panic buttons that they can press when they're in an emergency situation. The dispatch center can then locate the driver and send in police within a matter of seconds. Some cities—New York, for example—have passed legislation mandating the taxi companies install such security devices in their fleets to combat driver assaults.

School Buses

Interest in telematics applications for school buses has really taken off since September 11. As a result of the heightened level of security concerns in general in the U.S., more school districts are piloting wireless systems that can keep a central dispatch office appraised of a bus's location and status. In some cases, bus systems that serve disabled children are able to automatically give parents a five-minute warning via a phone ring in their homes so that children don't have to wait outdoors for the bus for extended periods of time. School districts and private bus companies are the ones shelling out the cash for these systems—parents pay in terms of increased school taxes.

"This is actually one of those markets where people don't mind paying a premium for the hardware and services," says Viquez.

There are also some systems being piloted that actually track children on the buses using radio frequency identification (RFID) tags. With these systems, a parent can get a confirmation that their child stepped onto or off of a particular bus.

Wireless data services provider GeoSpatial Technologies has been marketing these types of telematics products to school districts since 1999. The company now has four school districts using its tracking system

Mass Transit

Buses, trains and subways are obvious candidates for telematics applications—agencies have to manage a multitude of transportation vessels, often through city streets flooded with rush hour traffic. National systems such as Amtrak and Greyhound have had wireless tracking systems in place for a while. But recently, municipal transportation authorities have begun to pilot and deploy systems that allow them to track transit vehicles and re-route in cases of breakdowns, emergencies or special events.

Transportation agencies are seeing gains in terms of more efficient routing and dispatch, quicker response times in cases of service disruptions, and better overall service reliability. As of 2002, there were about 175 mass transit telematics systems in operation or under installation in the U.S., according to Viquez. Overseas, adoption has really taken off, with many cities integrating transit schedules with consumer applications that allow travelers to log on and determine when the next train or bus will arrive.

Emergency Vehicles

Post–September 11 security concerns have also increased emergency response agencies' interest in telematics. For a paramedic or firefighter, the ability to get navigation information and to transmit data about an incident to a central agency or a hospital can literally be live-saving. These institutions definitely need real-time connectivity, which can be difficult to maintain in remote areas.

In most cases, public-safety workers are using mobile communication units installed in their vehicles, but there's also a movement towards having handheld and laptop devices in hand once they get out of their vehicles. For police, these can help with background checks, while firefighters can use a handheld unit to bring up building plans as they're working on extinguishing a fire.

Also under trial are systems that integrate consumer and emergency vehicle telematics systems. For example, a consumer telematics system such as OnStar may become more sophisticated so that it can relay information about a crash to a paramedic as he's on his way to a crash site. The paramedic may be able to know whether the driver's airbag deployed, whether the vehicle rolled over, and perhaps even how many times it rolled over.

GeoSpatial Technologies implemented a telematics system for the Newport Beach Fire Department in 2001. When a fire incident is called in, the information is sent directly to mobile units in fire engines. A GPS system tracks the route to the incident and shows the firefighters where they're going. When a fire truck nears the incident, the mobile units bring up more information about the incident site—the city has digitized its building plans, so firefighters can view a map of a building as they approach it.

Maritime Industry

Ship crewmembers are using wireless data applications to communicate with and transmit information to on-shore colleagues. Today's technology allows at-sea workers to communicate in near real time with corporate headquarters or family members back on land, which opens up more possibilities in terms of wireless applications. More and more ship owners are taking advantage of satellite and digital radio communications as prices for airtime continue to drop. Companies such as Globe Wireless cater to the maritime market specifically, designing systems that allow ship owners to prioritize wireless data communications and make use of more economical transmission models. The upside for ship owners is the ability to get information to and from their ships faster, which makes getting up-to-date information for customers awaiting a shipment much easier.

Agricultural Industry

Though they're not who usually comes to mind when one mentions "mobile workers," farmers are actually extremely mobile. Most of them spend a good percentage of their working hours out in the fields, checking stock or sowing and harvesting crops. Large factory farms especially have taken up wireless systems to help them harvest and plant more efficiently. Farmers are using GPS systems for "precision farming"—allow them to get rid of the overlap as they sow, harvest and fertilize fields. Data is transmitted to mobile units on tractors out in the field. There's also a drive to get information to farmers while they're out in their field machinery—getting email and updates on grain prices, for example, is helpful.

Asset Tracking

Companies with valuable assets that need to be tracked are using location tracking technology to keep tabs on items as they're being shipped. Leasing companies are also finding location-based asset tracking useful—for example, John Deere can keep track of its construction equipment when it's out at job sites. The company can also use remote diagnostics to ensure that broken or malfunctioning equipment isn't used in the field.

Delivery Services

For many companies that focus on a daily delivery business model, the business changes every 24 hours. Newspaper subscribers go on vacation, new customers are added, or normal delivery times change. Wireless routing systems can take into account these changes and have a new delivery schedule waiting for drivers when they climb into their vehicles each morning. For example, Northside Forklift, a small company with a seven-truck fleet, is using a GPS service to coordinate its service vans and its billing system. Near real-time location data allows Northside to schedule forklift pickup and delivery according to traffic and weather conditions.

Field-force Workers

The value of a field technician's time is not in the vehicle, but rather on the job site. As little time spent in the vehicle is the goal—which reduces vehicle-related expenses and increases billable hours. So, for companies that dispatch workers in the field all day, a routing and tracking system that allows them to more efficiently service customers' needs is extremely helpful.

Car Dot Com

By Paul Lee

The vehicle will, over time, become an additional node on the network. This development is inexorable as all manner of vehicles will work better, increasing their value as a result of being connected. Trucks, cars, motorbikes, mechanical diggers, buses and any other vehicle form would all benefit from being able to transmit and receive mobile data.

Fanciful thinking and a relic of the dot-com era, or a misjudged anachronism? Deloitte Research argues that cellular mobile will gradually but increasingly become embedded into all vehicles. The incorporation of the technology will be used for a wealth of applications, from fleet management to driver security, from navigation services to maintenance.

Cellular mobile already plays a large role in the vehicle. One of the most common places to make a phone call is from a vehicle. In markets with a high volume of car usage, such as the U.S., it is estimated that over 10 percent of mobile calls are now made from vehicles. Generally, wherever both mobile networks exist, it is routine for both business and consumer users to call whilst in transit. And when traffic grinds to a halt, usage rises and the available cellular capacity is quickly sucked up by drivers and passengers looking to kill time. Despite the lamentably frequent instances of irresponsible usage of mobile phones when driving, the practice is here to stay.

The incorporation of mobile data capability in vehicles is still niche, but is steadily developing. We expect deployment will rise such that embedded mobile will be mainstream in vehicles in leading industrialized countries by the end of the decade.

High-end commercial and passenger vehicles already contain multiple, standalone networks. Each has a specific role. One key network will support driving functions, such as lights and, for certain models, also including the critical features of brakes, steering and acceleration. Another network may control content (audio, and, increasingly, video). A third network may hold information, primarily navigational data.

Mobile data will allow these networks to be connected together, so that vehicles evolve from being self-contained networks to being part of a larger, wider network. So whereas the engine-management system information system on a car is today part of an isolated network within the vehicle, over time the engine-management system will be able to communicate with the wide world, through it being connected via a mobile data connection.

Providing connections to these standalone networks generates value. Anyone who uses the Internet for any application, consumer or business, will concur with this view. Connectivity enhances the many information-based applications associated with vehicles.

A key application improved by mobile data is security. Both commercial and consumer markets can benefit. A well-known application for passenger cars is the automated emergency message. This is generated in the event of an airbag being deployed. The vehicle's coordinates are provided via Global Positioning System (GPS). The message can be sent in various formats, including Short Messaging System (SMS). The time saved in alerting emergency services can be life critical.

For business, an additional security application is fleet tracking. The growing value of commercial cargoes—the value of a pharmaceuticals or consumer electronics shipment can easily be in the millions of dollars—means that they are increasingly at risk of hijack. GPS data can be used to track a vehicle's route: any deviation from the expected route triggers an alarm. The immediacy of the alert can have a major impact upon the ability to retain the cargo and ensure the safety of the driver. Today this application is niche, due to cost, but over time, as it moves to becoming a shrink-wrap offering, deployment will become more commonplace.

One of the principle ongoing costs of vehicle ownership is maintenance; mobile data can help lower these and improve service levels. The trend for all vehicles is for increasing degrees of computerization, with even traditionally physical controls like braking and steering moving to a "drive by wire" approach. Engine management has for several years been controlled electronically. These trends are making vehicles more complex. This is reflected by the increasing reliance on computers in modern garages for both diagnostics and repairs.

Mobile data's key impact is that it can remove the need for onsite maintenance. Embedded mobile could allow a garage to remotely adjust fuel mix, monitor engine temperature or vary suspension. In a commercial environment, timely maintenance might mean a humidity-sensitive cargo arriving on time, rather than deteriorating on the roadside. For the consumer, it implies an even more reliable vehicle, improving satisfaction and engendering loyalty. For both consumers and commercial fleet owners, it can mean lower costs.

Embedded mobile can enhance the growing array of driver information services. The demand for one such service, traffic data, is rising in line with congestion growth. Currently, traffic information is usually relayed via a voice call. But variances in pronunciation can confuse the driver; key elements of the message might be missed. Mobile data would allow traffic data to be presented within a driver display, or even to be fed into a navigation system to generate an alternative route.

Mobile data can also be used to update the base data within navigations systems: presently this file becomes out of date as soon as the DVD is printed. In the medium term, customized information, such as news and weather bulletins, could be downloaded into a car's entertainment system in the early morning, then consumer on the journey into work.

Mobile data within vehicles can also be used for driver management. Fleet controllers can track of their vehicles' location, speed and hours of continuous usage. This data can improve fleet productivity. Car rental firms could be alerted of abusive use of vehicles or the car leaving its agreed boundary. Police forces may in the medium term be able to regulate car speeds if road conditions such as adverse weather conditions require it.

Much of the information currently generated by cars has commercial value which today is ignored, but which could be exploited via mobile data. Just two of the many data points—a car's speed and its location—could be collected. Aggregating such data from multiple vehicles would provide the backbone to a highly accurate traffic information service. The same two data points from a public bus could be used to provide to forecast arrival times to waiting passengers; the fleet manager could use this data to control spacing between vehicles.

Embedding cellular mobile into vehicles will become increasingly commonplace. In leading markets this functionality will have become mainstream by the end of the decade. Increasingly, vehicles both generate and depend on data; building in connectivity—which is only possible over a mobile network—raises the value of any type of vehicle.

Paul Lee is director of mobile and wireless at Deloitte Research. He is responsible for developing Deloitte's point of view on key trends in the mobile and wireless sector, with a specific focus on the impact of enterprise usage. He advises suppliers and users of mobile and wireless services and is a guest lecturer at London Business School.

CASE STUDY:
World Trade Center Uses Telematics for Cleanup

While the official end of the cleanup process at the site of the World Trade Center has not ended the grief and anger, there was at least a bit of positive news. The removal of 1.8 million tons of debris, predicted to take a year and cost $7 billion, finished two and a half months early, at a cost of $750 million.

The World Trade Center cleanup produced 108,342 truckloads of debris, which meant plenty of work for the numerous trucking companies that helped out with the cleanup process. But coordinating all the trucks and ensuring an efficient and safe disposal system was quite a task for New York City's Department of Design and Construction (DDC). Because the disaster site was treated as a crime scene, the city had to keep close tabs on all the loads that left the site, making sure that they ended up at the appointed salvage sites.

At first, the city had thousands of independent trucks just showing up at the cleanup site to work each day. "Every trucking company in the world showed up," says Yoram Shalmon, director of product management at wireless application service provider Power-LOC Technologies. "It's normal disaster mode." The DDC had been trying to manage everything using handwritten paper tickets that were given to truckers as they left the cleanup site. But, given handwriting errors, human fallibility and the easily destructible characteristics of a paper ticket, this system was not ideal.

There were also rumors that steel was being diverted from the appointed salvage sites, says Shalmon. Looking for a way to make the cleanup more efficient, the DDC called in consulting firms to help it review other technologies that might do a better job of managing the cleanup site. The DDC, with input from the Federal Emergency Management Agency (FEMA), local port authorities and law enforcement, decided that global positioning system (GPS) technology would be the best way to track and route the cleanup trucks.

The agencies worked together and decided to deploy a GPS system from PowerLOC Technologies, along with communications networks and on-site support and staffing from Criticom International, to manage the WTC cleanup site. Using GPS satellites, wireless technology, the Internet and a custom mapping system, PowerLOC set up a system that allowed city officials to see where each truck was at any appointed time. Each truck working at the site was fitted with a vehicle location device that monitored signals from 24 GPS satellites circling the earth. A signal providing the vehicle's exact location with its longitude and latitude was then sent via a wireless carrier to a central dispatcher.

People on the ground at the cleanup site would "open a ticket" each time a truck left the site by radioing the truck license plate, the GPS unit number, and what the truck was carrying. Criticom employees at the dispatch center could then monitor the flow of trucks from the WTC site to the locations designated to screen the materials. They were also able to set up the system so that they'd receive an alert if a truck went off route. "It was all crime evidence, so it all had to go through the FBI, and NYC police oversaw the site," says Shalmon.

PowerLOC used Cingular's Mobitex data network to transmit the satellite signals to a server, which then projected the measurements onto a map. The server would send commands such as "report your location every 10 minutes or every mile," ensuring that all users are tracked, and maintaining records in a database. A feature called GeoFence compared the satellite data to a virtual area that the user had designated to ensure that vehicles stayed on route.

The PowerLOC mapping system allowed dispatchers to track vehicle locations and watch for delays in the process. If a dispatcher noticed a bottleneck, he or she could redirect trucks to maintain the traffic flow. Likewise, if one truck was on its first trip, while all the others were already on their fourth or fifth, dispatchers could get in touch with trucking managers to find out what was going on with the straggler. This mapping information was also available to NYC officials via a secure Internet feed.

The system was also flexible enough to change along with the changes in the physical size of the cleanup site—as the city put blocks back into public use, the system had to adjust to accommodate the new map. Every day brought changes—whether it was new dimensions or new congestion areas.

The biggest challenge was getting the actual end users (the truck drivers, in this case) to consistently use the system. Shalmon says that there was a lot of griping among the truckers, who felt that they were being spied on. Then there was the intense pressure of trying to get the disaster site cleared as quickly as possible, and the feeling among the drivers that the system was slowing things down. Drivers who'd stop to check out before they left the site, only to find that their GPS units weren't functioning at that time, felt that the system was holding up the cleanup process and costing them time and money.

In the end, the system saved the DDC money and helped it finish the cleanup process in record time. Security was maintained and the agencies involved in the process now have digital records of the various steps involved in the cleanup.

THE NUTS AND BOLTS

Hardware: GPS in-truck units, wireless servers, PowerLOC tracker server

Applications: Camera monitoring and recording, vehicle tracking, Internet services

Number of users: 235 trucks

Cost of project: Not available

Network service providers: Cingular's Mobitex network

VARs/vendors: PowerLOC Technologies, Criticom International, Mobile Installing Technologies, Lynx PM

CASE STUDY:
Truckers Save Time with Location Tracking and Wireless Dispatching

Petro-Chemical Transport (PCT), based in Addison, Texas, hauls gasoline in locales all over the country. The company's drivers fill up their trucks at local terminals and make deliveries usually not more than 100 miles away. Then it's back to the gasoline terminal for another load. A driver's time is money in this industry—the more deliveries that a driver can make in a day, the more income for PCT.

The difference between PCT and most of its competition is that PCT operates its fleets from one central dispatch center, whereas most gasoline haulers have dispatch centers set up in each area where their drivers deliver. Up until PCT invested in a telematics system, this presented a major challenge: One dispatch center in Texas had to manage hundreds of deliveries all over the country every day.

Drivers would report to a terminal in their delivery area each morning and call PCT to let dispatch know that they were ready for work. Dispatch workers would then have to send a fax to the drivers with the details of their daily deliveries. Drivers would keep track of data such as deliveries, fuel usage and mileage information throughout the day, and then return to the terminal to file this information with PCT via fax at the end of the day.

If PCT needed to reach a driver to request an emergency delivery or change routes during the day, a dispatcher would have to page a few drivers and then wait for one to pull off the road and call the dispatch center on a pay phone. If a driver had an accident or an emergency, he was on his own, unless there was a phone nearby.

"It was always a challenge for us to communicate with all these trucks when they're often 1000 miles away," says Ron Ritchie, vice president of logistics at PCT.

Drivers would also mail in yet more documentation and then PCT workers would have to match everything up to make sure they had what they expected. The company employed a full staff of office workers to key in all the information—at least eight staffers were devoted to managing the faxing of delivery information each day.

"It was very inefficient—it took a lot of time, it took a lot of people, it just didn't work very well," says Ritchie.

In the early 1990s, PCT started looking into other options. The company's top priorities were that it wanted to be able to:

- Communicate with drivers at any time.

- Collect all the data they needed and transmit it back to the central dispatch center.

- Integrate seamlessly with their existing business systems so that it didn't have to be processed and re-keyed by office workers.

Ritchie says that the company looked into offerings from Qualcomm, which at that time was purely satellite-based communication services. He found that those products were great for tracking location information, but wouldn't help PCT keep track of mileage, fuel and delivery information. PCT also considered onboard computers that could give the company what it needed in terms of collecting truck information, but these didn't offer communication and tracking tools.

PCT finally settled on a system from Summary Systems that would transmit the information it needed to keep track of via cellular networks. PCT tested the system for about 60 days and found that it suited its needs.

"We had to first convince ourselves that you could reliably send data back and forth over cellular networks," says Ritchie. "Remember, this was the 90s, so cellular networks didn't offer as much coverage as they do nowadays."

Since then, PCT has added a GPS component (and also gone through a number of upgrades) to create a system that provides all the data that it needs. The company worked closely with its other systems vendors to ensure that the telematics applications integrated seamlessly into payroll, billing and other business programs.

Life for PCT and its drivers has gotten a lot easier since the telematics system deployment. Now, instead of drivers having to wait 20 to 30 minutes each morning and evening at a terminal to receive a faxed delivery schedule in the AM and to send back their logs in the PM, drivers just log on to their onboard computers each morning and receive the day's schedule immediately. PCT simultaneously receives confirmation that the driver has reported in.

As drivers go about their days, in-vehicle systems track information about the trucks (mileage, fuel consumption) and drivers enter delivery information into their in-vehicle computers. At end of day, drivers simply log off and their records are transmitted to PCT's server.

Drivers also now carry cell phones in their trucks, which makes it easier for PCT to reach them in cases of schedule changes. Drivers also appreciate the added security of having the phones in case of emergencies. Cellular coverage has improved to the point that it's not a concern for the company anymore, says Ritchie. "There are still a few areas where our drivers get intermittent coverage, but for the most part they're working in cities where coverage is available," he says.

Here's how the system works: At the time of the initial dispatch, PCT's central dispatch program creates a file with driver delivery information and places it on the company's server. The Summary Systems server then pulls files from PCT's UNIX box. When drivers sign on each morning, their in-vehicle computing units connect to the Summary Systems server and download the files. Work orders are transferred into truck computers. PCT then gets confirmation that the trucker showed up and got his work orders for the day. As the driver goes through his day, PCT can track him on GPS. At the end of day, when the driver signs off, PCT gets confirmation of the time for its own payroll records. PCT gets a file with billing data, payroll data, and driver vehicle condition reports. No one has to touch it—it's processed by the Summary Systems server and converted into ASCII files, then sent back to PCT's server. PCT's server sorts the data and populates the company's Oracle database and makes sure that all the data goes where it's supposed to.

The biggest benefit that PCT has seen from its telematics system is driver time savings. Instead of creating 30 minutes of "dead time" while drivers wait for their delivery schedule faxes each morning, drivers are already on the go making deliveries.

"We figure we're saving 30 minutes to an hour of truck time per shift, so that's up to two hours per day, which is 10 percent," says Ritchie. "That's a lot of money—it means you can do more work with the same amount of time."

PCT has at least doubled in size since its implementation, with 600 drivers currently hauling gasoline around the country, and the company has seen "huge" savings in overall costs, says Ritchie. The biggest one is driver payroll, since truckers are paid hourly. But the company has also decreased fuel consumption and required less support on the backend since eliminating its fax-based system.

The drivers were at first skeptical about the system, says Ritchie, but now they're not happy if for some reason they can't have a truck with a computer in it or their onboard computer goes down. The computers do a lot of their work for them, and the ability to communicate keeps them safe.

"I feel like this does give us a competitive advantage," says Ritchie. "We can operate cheaper on a central dispatch."

THE NUTS AND BOLTS

Hardware: Granite Communications Videopad

Applications: Scheduling, logistics, emergency response

Number of users: 700

Implementation time: Nine months

Cost of project: Not available

Network service provider: Verizon Wireless

VARs/vendors: Summary Systems

✪ AFTERWORD

If your company already has made the move to e-business, the extension to m-business can be a lot smoother than that initial move. The learning from the initial transformation is highly applicable to the task of mobilizing the company. An e-business has already identified critical information, revamped business processes and begun to analyze the new information stream of electronic data. Creating real-time, anywhere access to corporate data and systems by extending e-business to mobile devices is no longer a leap of faith, but simply a step forward from firm ground.

If your company is contemplating making the move from pencil and paper directly to mobile devices, you face different challenges. You may soon see the need to work backwards and automate more processes in the main office in order to benefit from the new data stream.

Both kinds of companies can benefit from their own and others' lessons at the School of Hard Knocks. After a period of hype and pipe dreams about the advantages of the always-connected business, enough real-world case studies are available, in this book and elsewhere, to illustrate the strategies and tactics that will pay off sooner rather than later.

The move to mobile is not a slam-dunk for any business. Strategists must carefully evaluate the opportunities for saving time, improving customer and partner satisfaction, and speeding cash flow.

After a serious assessment of the proposed mobile endeavor, careful planning and a step-by-step approach to implementation are the keys to success. Once an enterprise begins the mobile project, it must avoid "while we're at it" syndrome: adding extra features and applications that aren't must-haves. In the excitement of working with vendors—who are themselves excited at having clients— the enterprise can move too quickly and extend too far, creating a mess of less-than-compatible applications that are rejected by their intended users because they're too complicated.

Finally, the mobilizing company must keep in mind that the technology is changing so rapidly that a chosen platform may be considered legacy in just a few years. Building in flexibility by choosing open-architecture, device-independent applications is the safest route; however, this route may not offer the most options.

Whether you decide to move forward with a mobile or wireless rollout sooner or later, the time to start planning is now. Already, in your organization, employees are sneaking your company information and even your applications onto their personal mobile devices. Others are chafing to get mobile access to key business systems. It's up to you to take control and lead the charge.

✪ INDEX

Presumptive Design